Modeling, Identification and Simulation of Dynamical Systems

P. P. J. van den Bosch
A. C. van der Klauw

CRC Press
Boca Raton Ann Arbor London Tokyo

Library of Congress Cataloging-in-Publication Data

Bosch, P.P.J. van den (Paul P. J.)
    Modeling, identification, and simulation of dynamical systems / P.P.J. van den Bosch,
A.C. van der Klauw
        p.   cm.
    Includes bibliographical references and index.
    ISBN 0-8493-9181-4
    1. Dynamics.   2. Mathematical models.   3. System identification.
I. Klauw, A. C. van der (Alexander C.).   II. Title.
QA871.B697 1994
003′.85—dc20

94-19228
CIP

No claim to original U.S. Government works
International Standard Book Number 0-8493-9181-4
Library of Congress Card Number 94-19228
Printed in the United States of America 1 2 3 4 5 6 7 8 9 0
Printed on acid-free paper

Contents

Preface

This book gives an in-depth introduction in the areas of modeling, identification, simulation and optimization. These scientific topics play an increasingly dominant part in many engineering areas such as electrotechnology, mechanical engineering, aerospace and physics. Any engineering student needs some understanding of the general analysis and design tools for modeling and simulation. A thorough treatment of each of the topics of this book can be found in separate books. A unifying book that covers these topics is, however, scarce. In many engineering curricula these topics can not be treated in separate lectures. Consequently, we have developed over the past 10 years course material to present these important ideas in one lecture for undergraduate and graduate students. This book represents a unique and concise treatment of and the mutual interaction among the topics of this book.

Owing to their generally-applicable nature of modeling, identification, simulation and optimization, not too much background knowledge is required. It suffices if the student has some knowledge of dynamical systems, such as the mathematical models described via differential and difference equations, state models and frequency responses. This knowledge is acquired in the first two years of an engineering curriculum.

The contents of this book cover modeling and simulation of dynamical systems. These models can be derived by prior knowledge of the physical behavior of a system. Then white models arise that are based of the physical, electrotechnical, mechanical or chemical laws that describe the dynamical behavior among the variables of the system. If no or insufficient knowledge is present, a model can be based on measurements of the system. This way of obtaining a mathematical model is called black-box modeling or identification. Calculating dynamical time responses of these models is achieved with the aid of simulation tools. The chapter dealing with optimization discusses techniques for solving general nonlinear optimization problems as they arise in identification and many synthesis and design methods.

Chapter 2 emphasizes the main points in deriving mathematical models via prior knowledge concerning the physics describing a system. The first main point is to recognize that a system can and will have many different models, all describing different aspects of the same system. Consequently, a modeler has to be aware of the goal of using a model of his system, or his view on reality. The second main point is causality. A modeler has to recognize that causal models are artefacts. Reality don't bother about causality. All physical laws are formulated as relations among variables, without assigning in advance cause and effect. Bond graphs are a uniform approach for modeling systems as connection between physical components. The interaction between these components is defined as an exchange of power. This definition of power exchange allows us to isolate and assign the important variables in a model, namely the efforts and flows. Understanding these contributions of bond graphs allows a better ability in deriving models. It is shown that ordinary differential equations yield a powerful way of describing dynamical systems. Still, deriving differential equations is not always a trivial task. A more natural way is formulating a model via sets of acausal relations, not causal equations. This naturally leads to the use of DAEs (differential equations). The use of these DAEs are elucidated and explained. Still, there are many systems that need another description, namely by using difference equations, state events, discrete events. Also these mathematical models are discussed and the use of these for modeling is illustrated.

Identification of black-box models is discussed in Chapters 3, 4, 5 and 6. After a basic discussion dealing with statistics and stochastics, such as covariance and correlation, nonparametric and parametric models are discussed in describing the dynamical behavior of systems. It is shown that measurements of an unknown system can be used to approximate a linear model that describes the dynamical behavior of the system as good as possible. Parametric models such as FIR, ARX, ARMAX, OE and BJ models according to Ljung's nomenclature are introduced. Prediction error methods are discussed to define an appropriate optimization problem for finding the "best" set of parameters of a selected model that fits the measurements "as good as possible".

In practice, approximate identification is important, because it recognizes that reality is too complex to be modeled exactly. In Chapter 5 the use of approximate identification is elucidated. Chapter 6 discusses topics concerning the application of identification techniques in practice, such as input design, sample time selection, model structure and model order selection and prefilters. Moreover, validation techniques are discussed to judge whether a model really represents the dynamical behavior of a system as experienced in the measured data.

In Chapter 7 simulation is introduced. Simulation is often used in almost any engineering areas. It is shown that simulation is a powerful tool if properly used. Simulation is introduced as a numerical tool for calculating time responses of almost any mathematical model. The limitations of simulation are emphasized, because simulation programs seldom signalize the validity of their results. The numerical integration algorithms are illustrated, such as the influence of the step size or integration interval and the selection of a suitable integration method. In spite of many automatic step-size and integration-method selection algorithms, many nonlinear models require manual selection to obtain accurate and reliable results. As DAEs are becoming increasingly more important, the numerical behavior of DAE-solvers is illustrated.

Optimization is introduced in Chapter 8 as a generally-applicable tool for formulating and solving many engineering problems. It is stated that almost any engineering problem can be formulated as an optimization problem, such as finding the "best" solution among the many feasible solutions. Nowadays, powerful algorithms are available for solving these optimization problems. Still, a student has to be aware of the pitfalls of nonlinear optimization, such as its time-consuming, iterative nature and the possibility of locating a local instead of a global minimum.

It is our experience that the material of this book can be taught in a quarter or semester, accounting for about 100 hours of the student for attending the lectures, self study and examinations. We prefer an integral treatment of all four topics. However, sometimes one or two topics can be skipped depending on the availability of comparable lectures in the curriculum. This is especially true for the treatment of the numerical behavior of DAEs and approximate identification.

We greatly acknowledge the discussions with our colleagues and students at the Laboratory for Control Engineering at the Delft University of Technology. These have helped us to shape and complete this book. In particular, we have appreciated the contributions of Norbert Hofmeester, Ton van den Boom, Hans Visser and Ad van den Boom. The assistance of Barbara Cornelissen to improve our "American" text is gratefully recognized.
Finally we appreciate the indispensable background support of Tineke and Annelous, that has enabled us to finish the enterprise of writing a book in time.

Paul van den Bosch and Alexander van der Klauw

Eindhoven and Delft, March 1994.

chapter one

Introduction

In everyday life we encounter many interesting but complex processes. One can think of all kinds of applications, some of which are technical (chemical plants, trains, hydraulic installations, etc.), while others are biological, economic and physiological.

For many purposes, among them analysis and design, we want to be able to describe these processes in an understandable way. This means that we want to describe some aspects of a real-world object, the process, in an abstract way. We have to decide which characteristics to take into account, and which properties to ignore. It is the essence of the art of modeling to select only those characteristics, among the many available, which are necessary and sufficient to describe the process accurately enough according to the objectives of the modeler.

The construction of a valuable model requires a thorough understanding of the process under study and, additionally, of modeling techniques. In this book a survey is given of several modeling techniques which help us to derive a suitable model from existing knowledge. The modeling process consists of several consecutive steps, as illustrated in Figure 1.1.

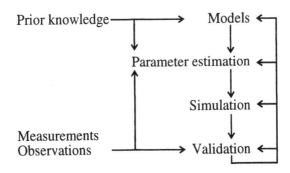

Figure 1.1 Modeling, parameter estimation and simulation

In Chapter 2 it will be shown that, based on a priori information of the physical laws that describe the basic dynamic characteristics of the process, a qualitative model can be derived. Such a model describes the relations between physical quantities.

Usually we want to have a quantitative model. However, in general our knowledge is not sufficient to know accurately all quantitative variables of such a qualitative model. Consequently, parameter estimation schemes have to be used to obtain proper values. Therefore a substantial part of this book is devoted to parameter estimation.

In Chapter 3 several discrete-time model representations are presented. Both parametric and nonparametric models are discussed.

Then, in Chapter 4, time- and frequency-domain analysis techniques are discussed to estimate the parameters of nonparametric models.
In Chapter 5 the identification of parametric models is considered. Prediction Error Methods are used for parameter estimation. These methods are based on the Least Squares method, which is treated thoroughly.

Some practical aspects of identification are discussed in Chapter 6.

One of the main goals of deriving models is to carry out experiments in an economic (i.e. fast, cheap, nondestructive, etc.) way. Simulation provides the tool for these experiments. A general treatment of simulation techniques is given in Chapter 7.

As will be shown, in parameter estimation and simulation it is necessary to solve nonlinear optimization problems. Chapter 8 contains a short survey of the procedures for their solution.

The present book provides an introduction to the interesting world of modeling, identification and simulation. It describes all phases in systems analysis and synthesis, in which models are involved, as illustrated in Figure 1.1. However, it has not been the intention to provide the reader with a thorough treatment of all aspects, as this would have taken several thousands of pages. After reading this book, the reader will find that many questions have been answered, but that some problems remain unsolved. The bibliography might then be of help in finding appropriate books specializing in that particular problem or field.

chapter two

Modeling

2.1 *Introduction*

Modeling is the art of creating mathematical descriptions of, e.g., physical, chemical or electrotechnical phenomena which appear in reality. These descriptions have to be relatively simple, yet accurate enough to serve the purpose of the modeler (Kheir, 1990).

It is important to recognize that many different models exist, all describing some parts of the same reality. It depends on our point of view and our intentions which part of that reality is described, as illustrated in Figure 2.1.

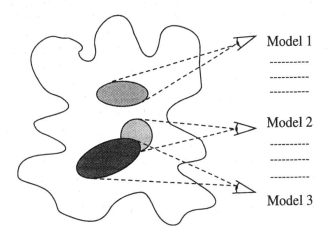

Figure 2.1 Different views yield different models

For example, an electronic resistor as physical component can be studied and described from many different viewpoints, e.g.:

- The resistor with resistance R [Ω] describes a linear relation between voltage u [V] and current i [A]: $u = R.i$.

- The resistor with temperature T [K], heat capacity C [J/K] and thermal conductance k [W/K] to the environment with temperature T_{env} [K] is described with $\dot{T} = [i^2 R - k(T - T_{env})]/C$.

- The resistor with mass m [kg] behaves as a mechanical body by $F_{\text{gravity}} = m \cdot g$.

- The resistor with inertia J [kgm^2] can be considered as a rotating mechanical body with a relation between torque M [Nm] and angular acceleration α [rad/s^2]: $M = J\alpha$.

- .....

In studying a chemical process, the modeler may be interested in the chemical reactions, in the physical working conditions, in the mechanical construction of the reactor, in the size and layout of the reactor, in the possible environmental impact of the reactor, in the financial return on investments, and in the dynamic behavior of the process. All these different views will yield different models. However, in this book we will restrict ourselves to models that describe the behavior of dynamic systems.

The essence of the art of modeling is to select only those characteristics, from among the many available, that are necessary and sufficient to describe the process accurately enough to suit the objectives of the modeler. This kind of abstraction from reality is a fundamental characteristic of science. The simplified models allow us to grasp the essentials from a turbulent and sometimes chaotic world. So, modeling is grasping the central issue from reality and translating it into an abstract language such as a mathematical model. Modeling is an essential part of all kinds of intellectual activities and enables us to understand, at least partially, reality. So, modeling is an integral part of all sciences and technologies ranging from philosophy and theology, via sociology, psychology, ecology and economy to civil, chemical and electrical engineering and physics.

A system is a collection of items that form a circumscribed sector of reality that is the focus of study. A system is a subjective entity with boundaries which include those items that are deemed most important to our objectives and exclude items of less importance. A model is the way we want to describe the salient features of the system under study. The model must possess some representation of the objects in the system and reflect the activities under which these entities act. So, a model is actually a reflection of the modeler's understanding of reality, its components and their interrelations.

An important decision in deriving models is the selection of the *system boundary*. This boundary determines which parts of reality, the process, will be taken into account. This isolated part of the process will be called the *system*. All parts of reality not belonging to the system are attributed to the *environment* of the system.

The selection of the system boundary can be critical. If too wide a boundary has been selected, it will become rather difficult to estimate the parameters of the model. It might even become impossible to analyze such a model and many important results will be hidden behind many irrelevant details. If too narrow a boundary has been selected, not all relevant aspects of the process will be incorporated in the model. This will yield unsatisfactory results and not meet the stated requirements.

Two main applications can be distinguished in using mathematical models. One consists of performing experiments in open loop (e.g., in making predictions), and the other of designing control systems which place the model in a closed loop. Making *predictions* requires an accurate description of the underlying physical laws. For instance, models used to forecast the weather or the future development of some economic variables, like unemployment, profit rates, etc, must be accurate. Small errors in the model will yield erroneous answers. In contrast, if the model is part of a *control system*, this control system will suppress disturbances, arriving from outside the system and errors resulting from modeling errors. So, a less accurate model can be used. Consequently, the requirements for accuracy depend on the application of the models.

In this chapter we will deal with some tools to assist in deriving appropriate models, among which the use of deduction, induction and of bond graphs. Still, modeling is more an art than a technique.

2.2 *Model building approaches*

Deduction

There are several ways of obtaining a model. First of all it is possible to derive a model by *deduction* based on prior knowledge of the process and insight into the behavior of dynamical systems. Deduction is used because it allows the application of general experience achieved from the modeling of other processes for a specific situation. This experience is very valuable. The prior knowledge of the process can be based on the *physical laws* that describe the mutual relations between the variables of the idealized process with idealized physical components. For example, if mass occurs in some model it will, in general, be assumed to be a point mass, without physical size, a flow will be considered to be laminar, concentrations are assumed to be homogeneously distributed in a tank, perfect mixing is presumed, etc. These physical laws can be described with the aid of algebraic and/or differential equations.

In general, much knowledge is available for defining the static, steady-state relations of processes. Based on this knowledge large chemical processes are characterized. These static models allow the calculation of operating points or steady state conditions which yields insight into the values of, for example, pressures, flows, temperatures or profits. In industry these static models are called flow sheets.

As soon as dynamics arise, differential equations are introduced. They describe a dynamic model. These equations are based on *laws of conservation* of energy, mass, momentum, etc. For example, the law of conservation of mass yields

$$\dot{m} = \Phi_{m,\text{in}} - \Phi_{m,\text{out}} \tag{2.1}$$

with m [kg] the mass and Φ_m [kg/s] the mass flow in and out, respectively. The conservation of energy states that

$$\dot{Q} = \Phi_{\text{in}} - \Phi_{\text{out}} \tag{2.2}$$

with Q [J] the amount of energy and Φ [W] the energy flow in and out, respectively.

Prior knowledge of the process can also offer additional information concerning the structure of the model. Not only the input-output relations between the variables, but also the structure of the model comes into focus. For example, if it is known that a system has two time constants, this knowledge can be used.

Out of all available structures a model structure is selected which guarantees the existence of two time constants or two real poles, instead of using a black-box second order model which may yield two complex poles. Prior knowledge of the presence, size and shape of nonlinearities can also be used, by adding them to the model description.

All previous strategies will yield a qualitative model. Sometimes it is also possible to predict the values of the parameters of the model rather accurately based on, for instance, the physical dimension of the process. All prior information concerning these parameters can be utilized by adding them to the model.

Induction

However, in general we do not have enough prior information to know accurately the values of all parameters. Then identification techniques are needed to find proper values. In the next chapters system identification techniques will be introduced which yield values of the parameters of a model, based on measured values of the input and output signals. These techniques can only be applied if some prior assumptions are made concerning, for instance, linearity, selection of input and output signals and system order. This way of obtaining information concerning the system is called *induction*. A selection criterion is required to decide which model is the "best" one given the measurements of the process.

White, gray and black models

Sometimes it is possible to derive a model by deduction only, based on the underlying physical laws and known parameters. Such a model is called a white-box model. In other situations almost no prior information is available, and the model has to be derived from the measured data of input and output signals, without any information concerning the internal structure and internal relations. These models are called black-box models. It is advisable to use as much prior information as possible. The amount available depends on the application area. In Figure 2.2 several application areas are shown, with an indication whether they normally yield white-, black- or gray-box models.

The sequence of actions ranging from modeling, estimation, simulation to validation has to be considered. They shape the activities necessary to derive suitable models. It can be recognized that an acceptable, fully-determined quantitative model is derived in several steps, as illustrated in Figure 2.3.

At any phase it is possible to go back one or more steps, because the results do not yield a model that satisfies the stated accuracy requirements. Validation is only possible if measurements or observations of the system are available.

Causality

One step in Figure 2.3 needs some explanation, namely the introduction of causality. Mathematical relations describe the mutual influence of several variables. For instance, Ohm's law defines the dependency between the current i and voltage u for resistor R by

$$\frac{u}{i} = R \qquad (2.3)$$

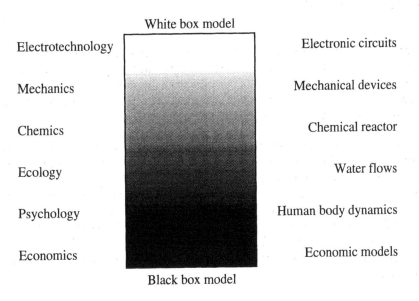

Figure 2.2 White-, gray- and black-box models

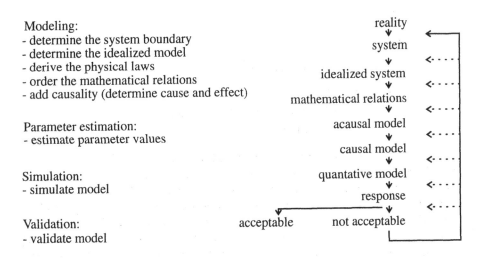

Figure 2.3 Iterative phases needed for modeling

The relation $u = R \cdot i$ suggests that a current i will introduce a voltage u. Still, the same relation is used to state that a voltage u will determine a current i through the resistor by

$$i = \frac{u}{R} \tag{2.4}$$

From a mathematical or a physical point of view, both equations are the same. They differ in the way cause and effect or input and output are assigned. Assigning cause and effect is called introducing *causality* into a model. Causality is introduced artificially in order to allow calculations. Causality is not physically determined ("reality doesn't bother about causality").

Remark In present-day simulation programs, only causal models can be dealt with. The model is defined by means of assignments instead of relations or equations. The relations or equations $u = i.R$ and $i = \frac{u}{R}$ are the same, the assignments $u = i.R$ and $i = u/R$ are different. Some, new simulation programs avoid the necessity of introducing causality. Then acausal models, models that only define the relations between their variables, are used. These acausal models can be described by acausal Differential Algebraic Equations (DAE) instead of causal Ordinary Differential Equations (ODE). So, adding causality is a necessity for using ODEs. ∎

Graphical representations

Graphical representations are an aid in describing models because they visualize the interaction between the variables. Several representations can be distinguished, ranging from those that are problem dependent with real components to those that are general purpose with a universal applicability:

- Drawings (artist's impression)

- Circuits

- Diagrams

    - Consecutive (flow chart, PERT)
    - Simultaneous

        * Causal (Block diagram / Bond graph)
        * Acausal (Bond graph)

Drawings give an artist's impression of the system. *Circuits* are closely related to one specific field of application with some standardization in the way that components have to be drawn. *Diagrams* represent a highly abstract way of describing dynamical systems. They are not coupled to a specific application field. Both consecutive and simultaneous diagrams can be distinguished. Examples of *consecutive* representations are flow charts utilized in programming and the PERT diagrams used by managers for planning purposes. In describing technical systems *simultaneous* diagrams are often used, especially *block diagrams*. It will be shown that the introduction of bond graphs can be useful. *Bond graphs* allow one first to draw the relations between variables and later, in a separate step, the required causality. In using block diagrams, both the relations and the causality have to be introduced at the same time.

2.3 Mathematical models

Several different types of models can be used to describe dynamical systems; these can be scale, mathematical or verbal models.

Scale models These permit examination of physical processes and are built on a reduced scale in order to study the real process at less cost. The wind tunnel for aerodynamic research (at NLR in Amsterdam), the towing tank for ship design (at MARIN in Wageningen) and the water research plant in Delft for simulating the water flow in rivers and estuaries are examples of scale models. In general, scale models are used for situations in which mathematical models are not accurate enough or cannot be calculated fast enough.

Mathematical models The dynamics of the system are described with the aid of (nonlinear) differential equations. This type of model is very flexible and will be dealt with later.

Verbal models The basic relations which describe the system are too complex and too little known to model them mathematically. Still, there is some understanding of the qualitative relations between the variables of the system. Examples of verbal models can be found in sociology and psychology.

The modeling techniques, discussed in this book, yield only mathematical models. These models can be subdivided into static and dynamic models and the dynamic models in turn can be divided into continuous and discrete models. These distinctions will be discussed at greater length.

2.3.1 Static models

A static model describes a (non)linear relation between the input and output of a function, as illustrated in Figure 2.4. Clearly, the output y is a function of both the parameters a and b and of the input u. For simplicity, the input and

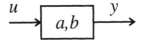

Figure 2.4 Static model

output are assumed to be scalars. We can make a distinction between linear and nonlinear functions. In Table 2.1 the combinations are elucidated.

In general, a model is said to be linear if its output is a linear function of its input. Still, linearity can also be defined on the basis of the relation between output and the parameters describing the function between input and output.

Processes are nonlinear. A correct model of a process will be saddled with nonlinearities. Our mathematical tools require linear models. A solution to this dilemma is the introduction of operating points. At a certain operating point the model can be linearized. Such a linearized model is suited for mathematical analysis and design. However, the answers will only be valid at the operating

Table 2.1
Nonlinearities

$y = F(u)$	$y = F(a, b)$	Example
linear	linear	$y = (a + b) \cdot u$
nonlinear	linear	$y = au + bu^3$
linear	nonlinear	$y = au/b$
nonlinear	nonlinear	$y = au/(bu + 1)$

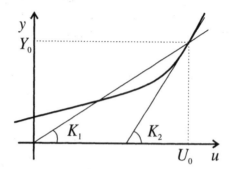

Figure 2.5 Nonlinear function $F(u)$

point under consideration. Each operating point will have its own, linearized model. In Figure 2.5 this linearization is illustrated.

Suppose (U_0, Y_0) represents the operating point. Then we have for small variations of Δu and Δy

$$u = U_0 + \Delta u \qquad y = Y_0 + \Delta y \tag{2.5}$$

Two different values of the gain can be distinguished, namely K_1 and K_2 with

$$K_1 = \frac{Y_0}{U_0} \qquad K_2 = \frac{\Delta y}{\Delta u} \tag{2.6}$$

In steady state, and hence at the operating point, we are interested in the gain K_1 represented by the relation

$$Y_0 = K_1 U_0 \tag{2.7}$$

For a linearized model the gain K_2 is more appropriate as expressed by the relation

$$\Delta y = K_2 \Delta u \tag{2.8}$$

Consequently, for small-signal analysis and for studying the stability of a model at an operating point the gain K_2 has to be selected. In designing processes, e.g. selecting temperatures, pressures, flows, and therefore for steady-state calculations, the gain K_1 has to be used. With small-signal analysis, which use only small values of Δu and Δy, the variables U_0 and Y_0 are assumed to be zero.

2.3.2 Dynamic continuous models

Several different mathematical models can be used for describing continuous or discrete dynamical systems:

Continuous models

- Distributed-parameter models (partial differential equations)
- Lumped-parameter models (ordinary differential equations)
- State-event models with state-dependent events

Discrete models

- Sampled-data models (difference equations)
- Discrete-event models with time-dependent events

These different model descriptions can coexist in one model, as illustrated in Figure 2.6.

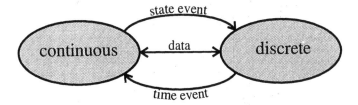

Figure 2.6 Interaction between continuous and discrete models

The interaction between the continuous and the discrete model is realized by means of state events or time events.

A *state event* is an event detected in the continuous model when a continuous variable passes through some threshold value, e.g. when a current i becomes zero or when a position x reaches a boundary. A state event can activate a switch in the continuous model or trigger some timing mechanism in the discrete model.

Time events are generated by timing processes inside the discrete event model. They can activate a switch or other actions inside the continuous model.

Distributed-parameter models

Since physical systems are continuous, they can, in general, be described adequately with the aid of either lumped- or distributed-parameter models. For example, describing the heat transfer through a wall is achieved by calculating the temperature gradient for all different materials of the wall. This process can be represented by a distributed parameter model, described with the aid of *partial differential equations*. Then, both the position x and the time t are independent variables of the model. For example, with $F(x, t) = 0$ the associated partial differential equation can become

$$\frac{\partial F}{\partial t} = a \frac{\partial F}{\partial x} \tag{2.9}$$

If the position-dependency can be neglected (lumped-parameter model, all activities are lumped together into one single point), just one or a finite number of differential equations come into focus. This simplification can be made by lumping all different materials into one homogeneous material with ideal characteristics, for example one concentrated mass with zero physical dimensions, one homogeneous surface, etc. If a finite number of positions is selected, the continuous position x is reduced into a finite number of values x_k. Each value x_k represents a model without position dependency.

This approach of selecting a finite number of "cells" or submodels is called the *Finite Element Method (FEM)*. Each cell has its own set of equations and variables. Consequently, the size of the model increases linearly with the number of cells. More cells yield more accurate results but also more calculation time is needed.

If the position dependency of the variables is removed, ordinary differential equations can be used. This conversion is illustrated in Figure 2.7.

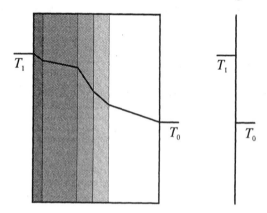

Figure 2.7 Conversion from distributed to lumped parameters

This simplification is important because the solution of partial differential equations by simulation is difficult and time-consuming.

Lumped-parameter models

Lumped-parameter models are described by *Ordinary Differential Equations* (ODE) or *Differential Algebraic Equations* (DAE).

- ODEs are formulated as explicit differential equations. Each derivative is defined at the right-hand side of an equation:

$$\dot{x}(t) = F\left(x(t), u(t)\right) \tag{2.10}$$

So, the value of $\dot{x}(t)$ can be calculated explicitly for each value of t.

- DAEs are formulated as implicit differential equations

$$F_1(\dot{x}(t), x(t), u(t)) = 0 \tag{2.11}$$

or as singular differential equations

$$E\dot{x}(t) = F_2(x(t), u(t)) \tag{2.12}$$

with E a singular matrix. E's singularity prohibits the calculation of E^{-1} and, consequently, (2.11) cannot be reformulated as an ODE. Also the presence of algebraic equations in variables, of which the derivatives are defined in an explicit differential equation, makes these equations implicit differential equations or DAEs.

Example 2.1 Suppose both $\dot{x}_1(t)$ and $\dot{x}_2(t)$ are defined explicitly, and $x_1(t)$ and $x_2(t)$ are related by an algebraic equation:

$$\dot{x}_1(t) = x_1(t) + 2x_2(t) + u(t) \tag{2.13a}$$

$$\dot{x}_2(t) = -x_2(t) - 3u(t) \tag{2.13b}$$

$$x_1(t) = x_2(t) + L \tag{2.13c}$$

This is a set of DAEs. Without additional effort, this example cannot be formulated as a set of ordinary differential equations. ⬜

The difference between ODE, stiff ODE and DAE is elucidated with the following example.

Example 2.2 Suppose two masses m_1 at position x and m_2 at position y at distance L are connected by a spring with spring constant k. The spring introduces a force $f(t)$ on both masses. This model is described as

$$m_1 \cdot \ddot{x}(t) = f(t) \tag{2.14a}$$

$$m_2 \cdot \ddot{y}(t) = -f(t) \tag{2.14b}$$

$$f(t) = k \cdot (L + y(t) - x(t)) \tag{2.14c}$$

This model is an ODE. For large values of k, the spring becomes stiff. The ODE becomes a stiff ODE, as discussed in Section 7.5. If the value of k increases even more, the connection between the two masses becomes rigid. The algebraic equation in (2.14c) is replaced by the algebraic constraint $x(t) - y(t) = L$. Then, the stiff ODE becomes a DAE. ⬜

Ordinary differential equations can be solved adequately with numerical integration methods, as discussed in Chapter 7. Differential algebraic equations introduce more numerical problems. The solution can be obtained by trying to reformulate the DAEs into a set of ODEs. This can be achieved by differentiation of the algebraic constraints, e.g. with symbolic formula manipulation. If the model is still described with DAEs, other methods have to be used, e.g., implicit numerical integration methods such as Gear's method (Gear, 1971 and Brenan, 1989), as discussed in Chapter 7. In Section 2.4 the index of a DAE is discussed.

State-event models

State-event models are continuous models that can generate discrete actions. At some time instants, when some continuous variable passes some threshold value, a discrete action is executed. For example, many gas-fired central heating units are equipped with a gas valve that can only be open or closed (no intermediate value). If this unit is controlled by means of a room thermostat, the unit will be switched on when temperature drops below some level and switched off if the measured temperature becomes larger than another, upper, limit. These switching actions are completely determined by the room temperature and the room thermostat. There is no synchronization of the switching time with any other process. Switching the central heating unit happens a few times an hour.

Another example deals with power electronic converters. These devices are able to convert, with almost no losses, large amounts of electric energy, e.g., from some power source to an electric motor. Low losses are achieved with the aid of electronic switches such as thyristors and transistors. These semiconductors can be partly controlled. A computer calculates the required points of time when a thyristor has to be fired. This point in time depends on the actual values of certain currents and voltages. A thyristor stops if the current or voltage drops below some threshold value. Again, this point in time is determined by the actual values of the variables in the simulation model. Switching on and off occurs very frequently.

In Chapter 7 the solution of continuous differential equations will be dealt with. It turns out that due to a numerical integration procedure, the variables are calculated at a limited number of points in time. The time period is called the integration interval.

A problem can arise if the switching times of the central heating unit or the thyristors do not coincide with these points in the timing of the numerical integration procedure. Due to the very long switching periods of a central heating unit (2 to 10 minutes) compared to an integration interval of, for example, 1 second, an acceptable error is made by bringing the switching time at the following integration time step, as illustrated in Figure 2.8. The switching time is adjusted to the integration step.

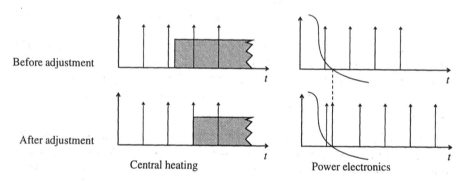

Figure 2.8 Synchronization of events and integration steps

This adjustment of the switching time is not allowed in the case of power converters. In contrast, the integration steps have to be adjusted to the switching time in order to maintain accuracy (Van den Bosch and Visser, 1990). The switching time has to be searched for by means of additional calculations. If it has been detected, the integration time step is adjusted, so that both coincide. Subsequently, the integration continues.

Synchronizing events to the integration procedure is executed automatically. The synchronization of the integration procedure to the events requires additional effort. General purpose simulation programs and simulation languages do not take care of this last type of synchronization, and dedicated attention has to be paid to realize it.

2.3.3 Dynamic discrete models

Sampled-data models

Sampled-data models arise when a computer has become part of a system, for example when a computer is used for controlling some process. The computer samples the process with the aid of AD-converters and controls the process by DA-converters. This sampling is done at fixed points in time. The time distance between two points is called the sampling period.

This type of model can easily be described with the aid of *difference equations*. Solution of difference equations with the aid of the digital computer is straightforward. The solution of the difference equation

$$y(k) - 1.5y(k-1) + 0.5y(k-2) = 4u(k-1) + u(k-2) \qquad (2.15)$$

using a programming language and assuming that the input $u(k)$ is stored in array u[k], is found by the following statements:

```
y1 = 0
y2 = 0
u1 = 0
u2 = 0
For k = 1 to 1000 do
    Begin
        y[k] = 1.5*y1 - 0.5*y2 + 4.*u1 + u2
        y2 = y1
        y1 = y
        u2 = u1
        u1 = u[k]
    End
End
```

Sampled-data models are characterized by a calculation sequence at fixed points in time. They can be mixed with continuous models, but then a synchronization between the sample time and the integration steps has to be realized. In general, the integration interval has to be selected smaller than the sample time to preserve the continuous behavior of the continuous part of the model.

Discrete-event models

Discrete-event or time-event models have no fixed sample period. All variables retain their values in between two events. At the occurrence of an event the variables may change their values. An event can happen due to external events or due to some timetable. This timetable can be produced as a result of internal activities. Examples of time-event models are queuing models of ships at sluices. traffic, people in queues at cash points in shops and the description of arrival, stay and departure of patients in hospitals or clinics. The time between two succeeding events can vary considerably, as illustrated in Figure 2.9.

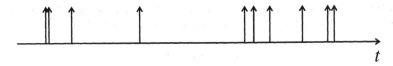

Figure 2.9 Unevenly distributed events of discrete-event models

In models described by means of differential or difference equations, variables are defined as functions of time t, e.g., $x(t)$ or $x(k)$. The state of the model is defined by the values of the variables x. The state x represents a physical quantity. The size of x indicates the amount of the physical quantity associated with x.

In models describing discrete-event systems, the variables represent timing processes, such as the arrival time at a queue or server, or the cycle time of a part of a production process. The state of a variable defines a point in time at which the variable will become active. The value of a state x represents timing information and NOT a physical quantity.

2.4 *Index of a DAE*

In this section, the structure of a DAE is analyzed. It will turn out in Chapter 7 that the ability to solve a DAE depends on its index. A DAE with index 0 or 1 can easily be solved. A DAE with index 2 requires advanced numerical procedures to achieve accurate results, while an index 3 DAE lacks, up to now, a reliable numerical solution procedure.

Suppose the variables of a model are formulated partly with differential equations and partly with algebraic equations

$$\dot{x}(t) = F[x(t), y(t), t)] \tag{2.16a}$$
$$G[x(t), y(t), t] = 0 \tag{2.16b}$$

Because $\dot{y}(t)$ is not explicitly defined, these equations shape a DAE. If we differentiate $G[x(t), y(t), t]$ once we obtain

$$G_x[x(t), y(t), t]\dot{x}(t) + G_y[x(t), y(t), t]\dot{y}(t) + G_t[x(t), y(t), t] = 0 \tag{2.17}$$

If $G_y(x(t), y(t), t)$ is not singular, an explicit equation for $\dot{y}(t)$ can be derived. Consequently, by differentiating the algebraic constraint or equation, this equation is reformulated as a differential equation. Then, the description of the model changes from DAE to ODE. If only one differentiation is necessary, hence $G_y[x(t), y(t), t]$ is not singular, the original model (2.16) has an index of 1. If $G_y[x(t), y(t), t]$ is singular, the equations have to be reshuffled and the resulting algebraic equation is again differentiated until all variables can be explicitly formulated. If two or more differentiations are required, a higher index results. The number of times differentiation is applied determines the index of a DAE.

Remark If no differentiation of a DAE is required, it is an ODE. Consequently, an ODE is a DAE with index 0. ▮

The effect of differentiation becomes evident from the following example.

Example 2.3 Suppose that two masses are connected by a rigid rod, so that the distance between the two masses is fixed by the length L of the rod:

$$x_1(t) - x_2(t) = L \tag{2.18}$$

If we differentiate this algebraic equation once, we obtain

$$\dot{x}_1(t) - \dot{x}_2(t) = 0 \tag{2.19}$$

Instead of formulating a constraint on the position, a constraint on the velocities is now stated. Theoretically, differentiation has not changed the relation between the two masses. If at $t = 0$, the distance is fixed at L, the constraint imposed on the two velocities guarantees that the distance will not change. Numerically, drift can be introduced in solving these equations. Another differentiation of this equation imposes constraints on the acceleration, namely $\ddot{x}_1(t) - \ddot{x}_2(t) = 0$.

$\square$

Nilpotency

In this section we will explain a more general approach for determining the index of a DAE. For simplicity we select a linear constant coefficient DAE.

$$A\dot{x} + Bx = f(t) \tag{2.20}$$

For this class an analytical solution of (2.20) can be found by decoupling it into a differential and an algebraic part. This decoupling and consequently the character of the solution is determined by the characteristics of the matrix pencil

$$\lambda A + B \tag{2.21}$$

The pencil is called *regular* if it is invertible ($\det(\lambda A + B) \neq 0$)) for some values of λ. Then, there exist nonsingular matrices P and Q transforming the pencil into the *Kronecker canonical form* (KCF).

$$PAQ = \begin{bmatrix} I & 0 \\ 0 & E \end{bmatrix} \qquad PBQ = \begin{bmatrix} C & 0 \\ 0 & I \end{bmatrix} \tag{2.22}$$

where E is a matrix in Jordan canonical form, having on its diagonal blocks $(E_1, .., E_k)$ each of them in the Jordan form

$$E_i = \begin{bmatrix} 0 & 1 & \dots & 0 \\ 0 & \dots & 1 & 0 \\ 0 & \dots & 0 & 1 \\ 0 & & & 0 \end{bmatrix} \tag{2.23}$$

It can be seen that for matrix E there exists an integer m such that $E^{m-1} \neq 0$ and $E^m = 0$, with m the size of the largest of its diagonal blocks E_i. Matrix E is called *nilpotent* and m is called its *degree of nilpotency*. The integer m is called also the index of matrix pencil (A, B).

Applying the above PQ transformation to the DAE (2.20) it is possible by scaling the equations by P and transforming the variables by matrix Q to decouple it into two completely independent subsystems

$$\dot{y}_1(t) + Cy_1(t) = g_1(t) \tag{2.24a}$$
$$E\dot{y}_2(t) + y_2(t) = g_2(t) \tag{2.24b}$$

where

$$Q^{-1}x(t) = \begin{bmatrix} y_1(t) \\ y_2(t) \end{bmatrix} \qquad Pf(t) = \begin{bmatrix} g_1(t) \\ g_2(t) \end{bmatrix} \tag{2.25}$$

The *differential part* (2.24a) is an ODE with the analytic solution

$$y_1(t) = e^{-Ct}y_1(0) + \int_0^t e^{C(s-t)}g_1(s)\,ds \tag{2.26}$$

Numerical methods exist for solving (2.24a).

The *algebraic part* (2.24b) consists of chains of differentiators corresponding to Jordan blocks E_i and is called a canonical nonstate subsystem. Its analytic solution can be derived by rewriting (2.24b) as

$$(E\frac{d}{dt} + I)y_2(t) = g_2(t) \tag{2.27}$$

giving

$$y_2(t) = (E\frac{d}{dt} + I)^{-1}g_2(t) = \sum_{i=0}^{\infty}(-1)^i \left(E\frac{d}{dt}\right)^i g_2(t) \tag{2.28}$$

Because of the time invariance of E we can move E in front of the operator d/dt and because of the m nilpotency of E we can cut the above sum at $i = m - 1$, so that

$$y_2(t) = \sum_{i=0}^{m-1}(-1)E^i g_2^{(i)}(t) \tag{2.29}$$

Comparing the solution (2.26) of the differential part (2.24a) with the solution (2.29) of the algebraic part (2.24b) we notice two important differences

- The solution (2.29) of the algebraic part (2.24b) involves derivatives up to order $m - 1$ of function $g(t)$.

- The solution (2.26) of the differential part (2.24a) allows arbitrary initial conditions $y_1(0)$ while consistent initial conditions $y_2(0)$ for the algebraic part (resulting in a smooth solution) are uniquely determined by (2.29).

The value m of the index of pencil (A,B) determines the character of the solution of (2.20). In the case of the constant coefficient DAE the index of the matrix pencil (A,B) (2.21) is called the nilpotency index of the corresponding DAE.

Remark In the above reasoning we have used the assumption of the existence of the *KCF* of the matrix pencil. This is equivalent to its regularity. If the pencil is *singular* (not regular) then the DAE (2.20) has none or an infinite number of solutions for a given initial value.   ∎

2.5 *Examples*

A number of examples illustrate the building process of models of dynamical systems. Deduction is used to derive these white-box models.

1 – Electronic circuit

If the components of an electronic circuit are known, it is possible to derive an accurate model by using deduction only. The electrotechnical laws are known and together with the values of the parameters of the components they yield a white model. In Figure 2.10 an electronic circuit is drawn, consisting of an inductance L [H], a capacitor C [F] and a resistor R [Ω].

This diagram defines the boundary of the system. The components, connected by electric wires, shape the system, which is idealized in that all parasitic

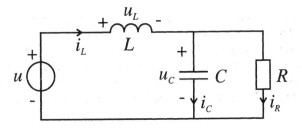

Figure 2.10 Electronic circuit

effects in the components and in the wires are neglected. The laws used in electronics yield the following mathematical relations

$$u = u_L + u_C$$
$$i_L = i_C + i_R$$
$$i_R = \frac{u_C}{R}$$
$$u_L = L\frac{di_L}{dt}$$
$$i_C = C\frac{du_C}{dt}$$

If u, R, L and C are given, the five unknown variables (u_L, u_C, i_L, i_C and i_R) can be solved with the aid of the five equations.

If we use integrators instead of differentiators, the model becomes as illustrated by the block diagram of Figure 2.11.

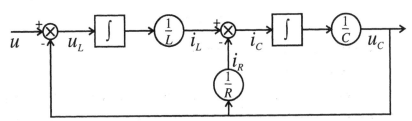

Figure 2.11 Block diagram of electronic circuit 2.10

Note that in drawing this block diagram we must introduce causality. We have to decide whether, for example, the current through the resistor determines the value of its voltage or vice versa. Because models are better suited for simulation if they do not contain differentiators, both the inductance and the capacitor are represented by their integral equation.

By defining u_C as output and u as input, and using the Laplace operator s, the transfer function $H(s)$ and the state equations become

$$H(s) = \frac{R}{RLCs^2 + Ls + R} \tag{2.30}$$

$$\begin{pmatrix} \dot{i}_L \\ \dot{u}_C \end{pmatrix} = \begin{pmatrix} 0 & \frac{-1}{L} \\ \frac{1}{C} & \frac{-1}{RC} \end{pmatrix} \begin{pmatrix} i_L \\ u_C \end{pmatrix} + \begin{pmatrix} \frac{1}{L} \\ 0 \end{pmatrix} u \tag{2.31}$$

This model, either represented as a block diagram or as a transfer function/state equation, has been obtained by using prior information only. No measurements have been used. Electronic circuits are examples of white-box models if the parasitic effects can be neglected and the values of the components are known sufficiently accurately. In practice, this is not always true owing to tolerances. Then some adjustable resistors are used to tune the circuit to its ultimate required behavior.

2 – Heating a house

Figure 2.12 shows a house that has a heating element which introduces an energy flow Φ_i [W] into the house. There is a flow of energy out of the house, indicated by Φ_o [W], as a consequence of the difference between the inside temperature T [K] and the temperature T_o [K] outside.

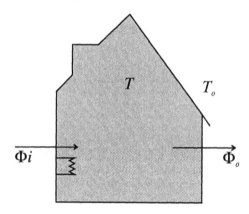

Figure 2.12 Heating a house

Two basic laws describe the temperature T

- conservation of energy;

- the size of Φ_o is proportional to the difference in temperature $(T - T_o)$.

The first law indicates that the difference between Φ_i and Φ_o is used to increase the energy Q [J] stored into the house. The relation between Q and the temperature T is given by the energy capacity of the construction of the house, namely C [J/K], or

$$Q = C \cdot T \qquad \text{[J]} \tag{2.32}$$

Taking the time derivative and noting that the net energy flow $(\Phi_i - \Phi_o)$ is the time derivative of Q, we have

$$\dot{Q} = (\dot{CT}) = C \cdot \dot{T} = \Phi_i - \Phi_o \qquad \text{[W]} \tag{2.33}$$

Moreover, Φ_o is proportional to the temperature difference in the walls

$$\Phi_o = k \cdot (T - T_o) \qquad \text{[W]} \tag{2.34}$$

with k [W/K] the energy conductivity of all walls and windows.
Taking both equations together, the required model becomes

$$\dot{T} = \frac{1}{C}(\Phi_i - k \cdot T + k \cdot T_o) \qquad \text{[K/s]} \tag{2.35}$$

Remark In contrast to the electronic circuit, the parameters of this model are not known in advance. In general they cannot be calculated either, because we have idealized and simplified the process of energy storage and energy transmission considerably. All walls are lumped together into one hypothetical wall where the energy transmission takes place. Moreover, no temperature drop through the wall is taken into account. In practice the temperature will drop in the wall (Figure 2.7) depending on the thickness and conductivity of the materials used for constructing it. Consequently, the parameter k of the model does not exist as a physical quantity in reality. The model represents a simplification of reality. The parameters can be approximated by tedious calculations or estimated by applying parameter estimation techniques to measurements of the temperature in the house. The latter approach is based on a black-box model of the house. No prior information is used, except for assumptions about linearity and order and the selection of inputs and outputs. ∎

Remark If the size of a model increases, the *dimensional analysis* can be used to detect some types of errors in the equations. This only concerns the physical units of the variables. The units of the variables on the left- and right-hand sides have to be the same. For example, a check of

$$Q = C \cdot T \qquad [\text{J}] \tag{2.36}$$

yields that Q is defined as [J] and $C.T$ as [J/K]·[K]=[J]. So, this relation satisfies the test. On the other hand, the equation

$$Q = k \cdot (T - T_0) \tag{2.37}$$

has to be wrong because the dimension Q [J] does not fit the dimension $k \cdot T$ = [W/K]·[K] = [W]. Dimensional analysis is both a simple and a useful tool for analyzing the relations between physical variables. The usefulness increases if units are assigned and written down for all physical variables and parameters. ∎

3 – Prey-predator model

In some isolated ecological environments the presence of rabbits (R) and foxes (F) can be assumed. The natural growth of isolated populations of either rabbits or foxes is proportional to their actual size. If sufficient food is available rabbits will increase. Without prey, foxes will decrease

$$\dot{R} = a.R \qquad \dot{F} = -b.F \tag{2.38}$$

Because rabbits can be considered as food for foxes, the models can be extended if both animals meet each other. The interaction is proportional to the sizes of the populations of rabbits and foxes

$$\dot{R} = a.R - c.R.F \qquad \dot{F} = -b.F + d.R.F \tag{2.39}$$

A further extension is possible by taking into account the competition between rabbits for food. This competition can be assumed to be proportional to the chance of meetings between two individual rabbits, which can be represented by R^2

$$\dot{R} = a.R - c.R.F - e.R^2 \qquad \dot{F} = -b.F + d.R.F \tag{2.40}$$

Again this model has been achieved by deduction. Measurements are necessary to estimate appropriate values for the parameters a, b, c, d and e. Solving this model with simulation techniques will yield results as shown in Figure 2.13.

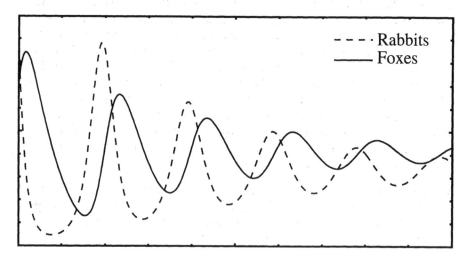

Figure 2.13 Prey-predator relations

4 – Chemical reactions

Modeling chemical reactions requires understanding the kinetics of a reaction. These kinetics are based on the concentration of the chemical elements in a reactor. Suppose two materials A and B react with each other as shown in Figure 2.14.

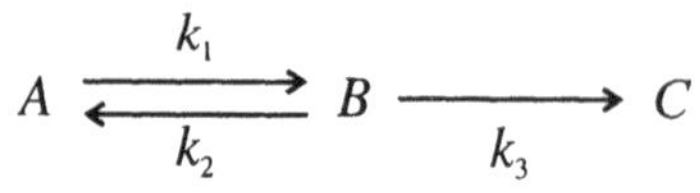

$$A \underset{k_2}{\overset{k_1}{\rightleftharpoons}} B \xrightarrow{\ \ k_3\ \ } C$$

Figure 2.14 Chemical reaction

The concentrations of A, B and C are given by x_A, x_B and x_C. The reaction from A to B has a dynamic constant k_1, from B to A a dynamic constant k_2 and from B to C k_3. The basic dynamic relations follow from

$$\dot{x}_A = -k_1 x_A + k_2 x_B$$
$$\dot{x}_B = k_1 x_A - k_2 x_B - k_3 x_B$$
$$\dot{x}_C = +k_3 x_B$$

Additional information concerning the volume makes it possible to calculate the masses of the different components.

5 – Bouncing ball

When a ball is released from some initial height x_0, the height x [m] can be calculated. First, the ball of mass m [kg], will fall owing to gravity g [m/s^2].

Second, if the ball with radius r [m] touches the floor ($x = r$), the ball will also act as a mass-spring process with spring constant k [N/m] and damping f [Ns/m]. A linear spring is assumed to describe this mass-spring process. Define v [m/s] as the velocity of the ball with a positive value for upward movements. Two different models are distinguished.
If $x > r$

$$\dot{x}(t) = v(t) \tag{2.41a}$$

$$\dot{v}(t) = -g \tag{2.41b}$$

If $x < r$

$$\dot{x}(t) = v(t) \tag{2.42a}$$

$$\dot{v}(t) = -g - \frac{1}{m}(f \cdot x - k \cdot (x - r)) \tag{2.42b}$$

With the initial conditions $x(0) = x_0$ and $v(0) = 0$, these two sets of equations describe a bouncing ball.

In the simulation program PSI/c such a model can be described as

```
height0 = .9;          % Initial height of the ball  [m]
g        = -10;        % Gravity constant            [m/s^2]
damping  = 10;         % Damping of ball             [Ns/m]
k        = 1e4;        % Spring constant of the ball [N/m]
mass     = 1;          % Mass of the ball            [kg]
radius   = .1;         % Radius of ball              [m]

height = INT(speed PAR: height0);   % Height of ball  [m]
speed  = INT(g-accel PAR: .0);      % Speed of ball   [m/s]
   % The acceleration is calculated as a consequence of the
   % spring-damper
accel = (speed*damping+(height-radius)*k)*(height<=radius)/mass;
   % [m/s^2]
```

Remark The accuracy of this model is improved, if the point in time the ball reaches and leaves the floor is detected by a state event. So, a state event has to be generated if $x = r$. ▌

6 – Level process

In a tank of area A [m^2] a liquid is stored. There is a flow Φ_o [m^3/s] of liquid out of the tank. The flow Φ_i [m^3/s] into the tank is controlled in a rather rough way. If the level x [m] becomes higher than x_{max}, the flow into the tank is stopped by means of a valve.
After three seconds the flow into the tank continues with a value Φ_{max}. This model is described as a differential equation, extended with a state event to detect the $x = x_{max}$ and one discrete event process to take into account the timing of closing and opening the valve in the flow into the tank.

$$\dot{x}(t) = (\Phi_i(t) - \Phi_o(t))/A \tag{2.43a}$$

$$\Phi_i(t) = \text{valve} \times \Phi_{max} \tag{2.43b}$$

Process 1 = state event to detect ($x = x_{max}$)
Process 2 = discrete event, triggered by process 1 and taking 3 seconds.

Process 1 closes the valve (valve=0) and process 2 opens the valve (valve=1). In the simulation program PSI/c such a model can be described as

```
area       = 2.;
f_max      = 3.;
flow_out   = 1;
level_max  = 12.;

% Differential equation to describe the level
level      = INT ((flow_in-flow_out)/area PAR: 11.);
    % level
flow_in    = valve * f_max;
    % flow into
valve      = FFL (0, process_1, process_2 PAR: 1, 0);
    % valve for flow_in

% Two processes.
% Process_1 is triggered if level >= level_max.
% It closes the valve.
% Process_2 is triggered 3 seconds after process_1 has been
% activated. It opens the valve.

process_1 = STE (level-level_max          PAR: 0.01, 1);
    % State-event
process_2 = DEV (process_1, 3             PAR: 0, 0);
    % Discrete event
```

2.6 *Bond graphs*

2.6.1   *Variables*

Besides the more often used block diagrams and circuit diagrams, bond graphs (Karnopp, Margolis and Rosenberg, 1990) play an important role in modeling physical systems in a systematic way. A bond graph is composed of components that exchange energy or power through connections. These connections are (power) bonds. Essential to this description are the components and the bonds. In contrast with block diagrams these connections represent an exchange of power instead of a one-way transport of information. Later a more detailed comparison between bond graphs, block diagrams and circuits will be made. First a description of bond graphs is given.

Each connection or *bond* transports power or an energy flux P. This power is the product of two variables, namely a variable called *effort e* and a variable called *flow f*. The effort e and flow f are generalizations of similar physical phenomena, as illustrated in Table 2.2.

This brief example emphasizes the potential of bond graphs. The same description and variables can be utilized in all different kinds of physical systems, such as electrotechnical, mechanical (translation and rotation), hydraulic, thermodynamical and thermal systems. If a physical system consists of an electromotor that runs a fluid pump, both components and variables of the different fields of electronics, electrotechnology, mechanics and hydraulics are merged in

Table 2.2
Similar physical phenomena

	Electrical	Mechanical	Hydraulic
effort e	voltage u	force F	pressure p
flow f	current i	velocity v	flow ϕ

one system description. Bond graphs allow a universal treatment of all these different application fields into one model.

Besides the generalized variables effort e and flow f, two other generalized variables can be distinguished, namely the displacement q and the momentum p, with q the time integral of f and p the time integral of e, so that

$$q = \int f dt \quad p = \int e dt \tag{2.44}$$

These variables q and p preserve their value among coordinate transformations and are called *preserved variables*. In contrast, the generalized variables e and f can change among coordinate transformations. In Table 2.3 these generalized variables are illustrated for several application fields. The generalized momentum only has meaning in the field of electrotechnology (pulse [Vs]), translation (momentum p [Ns]) and rotation (momentum L [Nms]).

Table 2.3
Generalized variables in physical systems

Application	Effort e	Flow f	Displacement q
Electric	Voltage u [V]	Current i [A]	Charge q [As]
Translation	Force F [N]	Velocity v [m/s]	Displacement x [m]
Rotation	Torque M [Nm]	Velocity ω [rad/s]	Angle ϕ [rad]
Hydraulics	Pressure p [N/m^2]	Volume flow Φ_v [m^3/s]	Volume V [m^3]
Thermodynamics	Temperature T [K]	Entropy flow $\dot{S}$ [W/K]	Entropy S [J/K]
Thermal	Temperature T [K]	Energy flow Φ [W]	Energy Q [J]

The mutual relations among these four generalized variables e, f, p and q and the three generalized elements R (resistor), C (capacity) and I (inertia) are illustrated in Figure 2.15.

2.6.2   Bonds

With the introduction of the variables effort and flow, the bonds can be defined more precisely by adding a direction and causality and by introducing the one-way bond, the so-called activated bond.

A bond is represented by a line. The symbol for the effort is put above or right of the line and the symbol for the flow under or left of the line, so a bond becomes as in Figure 2.16.

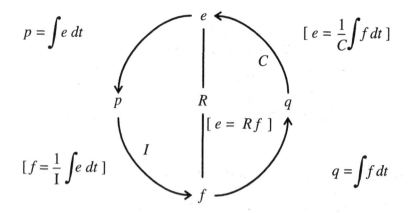

$$p = \int e\, dt \qquad\qquad\qquad [\, e = \frac{1}{C}\!\int f\, dt \,]$$

$$[\, f = \frac{1}{I}\int e\, dt \,] \qquad\qquad\qquad q = \int f\, dt$$

$$[\, e = R f \,]$$

Figure 2.15 Mutual relations among e, f, p and q

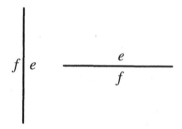

Figure 2.16 The bond

Direction

The direction of the positive power flow is indicated by a half arrow. The
half arrow is used to distinguish it from the full arrow which is used in block
diagrams to represent a transport of information. The direction of a bond is
illustrated in Figure 2.17.

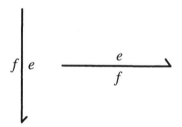

Figure 2.17 Direction of the bond

Causality

A second important characteristic of a bond determines the causality. In the
previous drawings of the bonds, no input or output can be detected, so these
bonds are acausal. Causality is introduced artificially to calculate the model
described by a bond graph. Physical processes do not possess causality. They
represent a parallel system with mutually influencing elements. Consequently,

causality means assigning cause and effect or input and output to a physical phenomenon as expressed by a mathematical expression.

Causality can be expressed in a bond graph by assigning a cross stroke at one of the two ends of a bond. If the effort e is the cause and f the effect (e.g., $f = G.e$), then the cross stroke is located at the component side of the bond. If the flow f is the cause and e the effect (e.g., $e = R.f$), then the cross stroke is located opposite to the component side of the bond, as illustrated in Figure 2.18.

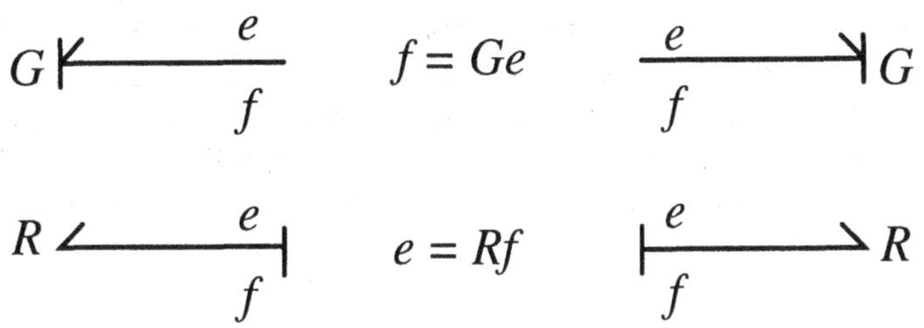

$$G \vdash\!\!\frac{e}{f}\qquad f = Ge \qquad \frac{e}{f}\!\!\dashv G$$

$$R \angle\!\!\frac{e}{f}\!\!\dashv \qquad e = Rf \qquad \vdash\!\!\frac{e}{f}\!\!\searrow R$$

Figure 2.18 Causality of R and G element

In locating the cross stroke it can be useful to notice that, as illustrated in Figure 2.19, the arrow indicating the "direction" of e is pointing towards the causal cross stroke. With the G element e "enters" this element and f is the effect, while with the R element f is the cause and e "leaves" the R-element. Hence, the effort e always points to the cross stroke.

Figure 2.19 Determining causality

Remark The causal cross stroke is introduced solely for determining the calculation sequence. It has no relation to the half arrow of the bond, which indicates the positive flow of power (where the modeler can choose the positive direction). ∎

Activated bonds

A characteristic of bonds is that they describe the mutual interaction between physical variables. Consequently, these variables will influence each other. For example, the load of an electromotor will influence the dynamic behavior of this motor. Although a direction and causality have been introduced, this return action of the variables still exists in a bond graph.

Sometimes there is no return action. For example, a controller will steer an actuator, e.g. a valve. The electronic design of the controller will be such that the return action of the valve on the controller electronics can be neglected. If there is no return action, an activated bond is introduced. As in the block diagram, it represents the flow of information and it is represented as a line with a full arrow as illustrated in Figure 2.20.

Figure 2.20 Activated bond

2.6.3   *Bond graph elements*

A bond has been defined as a connection between two elements. In this section
a description of the basic elements of bond graphs will be discussed. We can
distinguish sources, one-ports, two-ports and multi-ports. In Table 2.4 these
different elements are illustrated and defined. In fact the names of the electronic
components are selected for describing the generalized elements of bond graphs.
Still, a resistor can be any physical component with a linear relation between
effort and flow, for instance friction in fluid flows, or the consequence of a
translation or rotation.

Table 2.4
Bond graph elements

Symbol	Name	Function	Example
Sources:			
$S_e \dfrac{e}{f}$	e-source	e	u-source
$S_f \dfrac{e}{f}$	f-source	f	i-source
One-ports:			
$\dfrac{e}{f} \, R$	Resistor	$e = R.f$	Resistor
$\dfrac{e}{f} \, G$	Conductivity	$f = G.e$	Resistor
$\dfrac{e}{f} \, C$	Capacity	$e = \frac{1}{C} \int f.dt$	Capacitor
$\dfrac{e}{f} \, I$	Inertia	$f = \frac{1}{I} \int e.dt$	Coil
Two-ports:			
$\dfrac{e_1}{f_1} \, \text{TF} \, \dfrac{e_2}{f_2}$	Transformer	$e_2 = m.e_1$ $f_1 = m.f_2$	Transformer
$\dfrac{e_1}{f_1} \, \text{GY} \, \dfrac{e_2}{f_2}$	Gyrator	$e_2 = m.f_1$ $e_1 = m.f_2$	DC-motor
Multi-ports:			
$f_3 \mid e_3$ $\dfrac{e_1}{f_1} \, 0 \, \dfrac{e_2}{f_2}$	0-junction	$e_1 = e_2 = e_3$ $f_1 + f_2 + f_3 = 0$	Parallel
$f_3 \mid e_3$ $\dfrac{e_1}{f_1} \, 1 \, \dfrac{e_2}{f_2}$	1-junction	$f_1 = f_2 = f_3$ $e_1 + e_2 + e_3 = 0$	Serial

One-port

The *one-ports* capacity C and inertia I shape storage elements. Energy is stored and retrieved from these elements. There is no energy dissipation. Only in the R element energy is dissipated and withdrawn from the system.

The value of an element is mentioned after the symbols R, C or I. If an element has size R_1, C_i or L_j, the texts with the bond graph elements become $R : R_1$, $C : C_i$ and $I : L_j$ respectively.

Two-port

The *two-port elements* transformer TF and gyrator GY transport, without losses, power from one of their ports to the other one. Consequently, the product $e \cdot f$ of both ports is equal. With the variable m, the size of the effort and flow is influenced, not the power flow. The domain of these two ports can be different. For example, a gyrator is used to model a DC motor or a loudspeaker. One port is defined with electrical variables (u and i), while the other port uses mechanical variables, namely rotational torque M and angular velocity ω with the DC motor and translational force F and velocity v with the speaker.

DC-motor: (electrical and mechanical (rotation))

$$M = m.i \qquad\qquad (2.45a)$$
$$u = m.\omega \qquad\qquad (2.45b)$$

Loudspeaker: (electrical and mechanical (translation))

$$F = m.i \qquad\qquad (2.46a)$$
$$u = m.v \qquad\qquad (2.46b)$$

Multi-port

The *multi-ports* shape the junctions of bond graphs. The 0-junction represents a parallel circuit of elements. With a 0-junction all efforts are equal. The sum of the flows is zero.

The 1-junction shapes a serial connection of elements. Then all flows are equal and the sum of the efforts is zero.

Remark If a connection of different elements is detected in a bond graph a junction is present. The easiest way to decide whether this junction is a 0-junction or a 1-junction is to verify whether all efforts at the elements of the junction are equal (0-junction) or all flows have the same value (1-junction). ∎

2.6.4 Causality

In section 2.2 causality was introduced. Causality is not a natural fact that can be derived from the physical properties of components or variables. Bonds are connections between elements which influence each other. Causality is selected and added artificially to execute calculations. There are some rules for determining causality in a bond graph.

Source causality

Sources always prescribe causality. An effort source S_e will impose an effort on the system. Consequently, the cross stroke which indicates causality has to be drawn at the end of the bond where the effort e points to. A flow-source S_f imposes a flow on the system. Hence the cross stroke must be located at the component side S_f, since the flow "leaves" the source, and the effort enters the source. Both source causalities are shown in Figure 2.21.

Figure 2.21 Source causality

Integral causality

Although a capacitor or an inertia can be described both by means of an integral equation and a differential equation, the solution of these elements requires an integral equation. In using simulation software, an integrator is supported while a differentiator cannot be realized. This statement makes the effort the "input" of an inertia I, and the flow the "input" of a capacity C. Consequently, the cross stroke which indicates causality has to be drawn as illustrated in Figure 2.22.

$$f = \frac{1}{I}\int e\,dt$$

$$e = \frac{1}{C}\int f\,dt$$

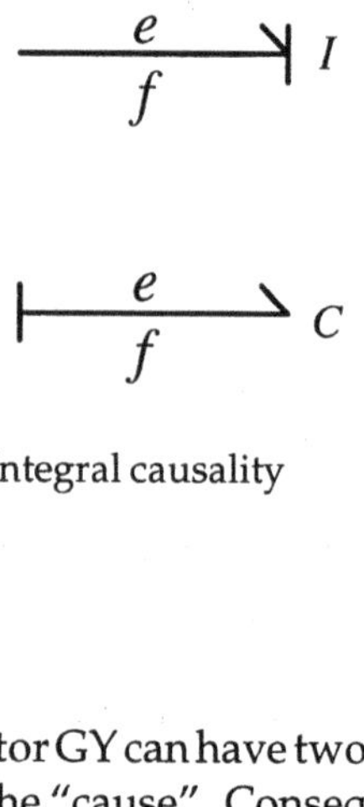

Figure 2.22 Integral causality

Causality of two-ports

The two-ports transformer TF and gyrator GY can have two different realizations of causality. In a bond either e_i or f_i is the "cause". Consequently, f_i or e_i will be the effect. Suppose e_1 is the cause of a transformer, then both e_2 ($e_2 = m.e_1$) and f_1 will be the effect and, subsequently f_2 will also be the cause. If e_1 is selected to be an effect of the cause e_2, then f_1 will be the cause and f_2 the effect. Both realizations of causality of a transformer are shown in Figure 2.23.

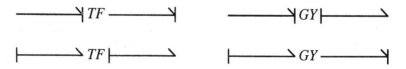

Figure 2.23 Causality of two-ports

Suppose e_1 is the cause of a gyrator, then both f_1 and f_2 ($f_2 = \frac{e_1}{m}$) are the effects and, subsequently, e_2 will also be the cause. If e_1 is selected to be an effect of f_1 ($e_2 = m.f_1$), so is e_2, and both f_1 and f_2 are the cause. Both realizations of causality of a gyrator are shown in Figure 2.23.

Causality with junctions

A 0-junction represents a parallel circuit with all efforts having the same value. Consequently, only one effort can be an "input". For the other bonds at the same 0-junction flows have to be "inputs". There is only one cross stroke at a 0-junction.

A 1-junction represents a serial circuit with all flows having the same value. This implies that only one flow can be an "input". The efforts in the other bonds of that junction will become "inputs". With the exception of only one (1) bond, all other bonds will have a cross stroke at a 1-junction. Examples of this causality associated with junctions are shown in Figure 2.24.

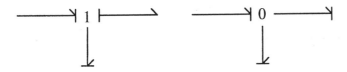

Figure 2.24 Causality with junctions

Causality of resistors and conductivities

The causality of the R and G element has been introduced in Figure 2.18, and is repeated in Figure 2.25. In practice, often either an R or a G element can be

$$f = G e \qquad \dfrac{e}{f} \quad G$$

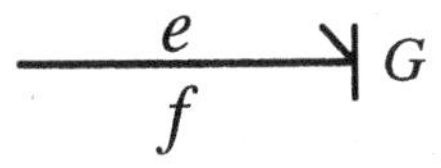

$$e = R f \qquad \dfrac{e}{f} \quad R$$

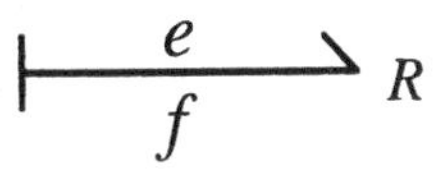

Figure 2.25 Causality of R and G elements

selected, depending upon the causality requirements.

2.7 *Comparison circuits, bond graphs and block diagrams*

In this section we will deal with the characteristics and the advantages and disadvantages of model representations based on circuit diagrams, bond graphs and block diagrams. Examples of these three different model representations are given in Figure 2.26.

The following observations can be made.

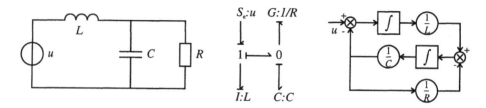

Figure 2.26 Circuit, bond graph and block diagram

- In a circuit, the variables represent *physical quantities,* in bond graphs *energy flows* and in a block diagram *signals*. Signals can be separated into two or more branches with still the same value of the signal on these new branches. In using circuits or bond graphs each split will introduce a division of the physical quantity among the new branches. The values before and after the split will differ, as illustrated in Figure 2.27.

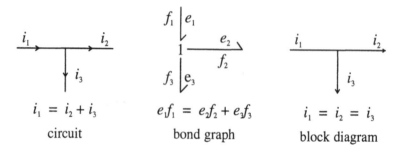

$$i_1 = i_2 + i_3 \qquad\qquad e_1f_1 = e_2f_2 + e_3f_3 \qquad\qquad i_1 = i_2 = i_3$$

circuit   bond graph   block diagram

Figure 2.27 Splitting into branches

- In circuits each connecting line represents one physical quantity which has influence in both directions. In bond graphs a bond represents two (2) variables, namely an effort and a flow. Again these variables have their influence on both adjacent components. In block diagrams a line represents just one (1) signal. This signal is fully determined by the block which calculates this signal. Blocks which have this signal as input can not influence the value of this signal. Consequently, the influence is only in one direction. In contrast with circuits and bond graphs a block in a block diagram always has an input and an output.

- In a circuit the elements are selected as problem-dependent, idealized physical *components*. The elements in a bond graph are general-purpose, idealized components while in a block diagram the elements represent *functions* between input and output.

- Due to the lack of an implicit *return action* between connected elements (blocks) in a block diagram, the physical return action has to be modeled with the aid of an additional feedback loop. Both circuit and bond graph do not need such an artificial extra loop. The return action is implied in the mutual influence of two connecting elements.

- A circuit is drawn with the symbols and variables of a specific problem domain such as electronics, hydraulics, mechanics, thermodynamics, etc.

This facilitates a close connection with the process to be modeled. However, the modeling of a process consisting of parts from different problem fields becomes difficult or even impossible with a circuit as model representation. Then, both a bond graph and a block diagram yield a more general approach.

- A circuit represents the physical relations between the components. No causality is added. In block diagrams causality always has to be included because all blocks have at least an input and an output. In bond graphs causality can be added. Without causality it represents the physical process with all its interacting elements. If calculations are required causality is a prerequisite and has to be added artificially.

- Because causality is added artificially, model representations with causality will, in general, not be unique. There can be many different ways for defining causality, which implies that there can be many solutions and, consequently, many different model representations. Of course, different causal model representations of the same acausal model describe the same dynamic behavior.

In Table 2.5 the characteristics of circuits, bond graphs and block diagrams have been put together.

Based on this comparison it can be concluded that a model representation

Table 2.5
Comparison Circuit, Bond graph and Block diagram

	Circuit	Bond graph	Block diagram
Character variables	physical	energy flow	signal
- number	1: physical	2: e and f	1: signal
- direction	2 directions	2 directions	1 direction
Element	component	component	function
Return action	included	included	extra feedback
Problem	dependent	general	general
Causality	acausal	acausal or causal	causal
Uniqueness	unique	acausal: unique causal: nonunique	nonunique

by means of a circuit has the disadvantage of being problem-dependent and the block diagram has the artificial addition of feedback loops to represent the return action. Moreover, by artificially adding causality a bond graph may become nonunique and a block diagram is nonunique. The freedom of assigning causality may lead to different models. On the other hand, causal models are necessary if these models are used for simulation purposes. So, causality may be a required characteristic.

2.8 *Examples*

1 – Electronic circuits

In the examples in Figure 2.28 a number of electronic circuits are described, and subsequently the associated bond graph is derived. Each circuit consists of either a voltage or a current source with two capacitors or inductances in parallel or in series.

Only a causal bond graph yields a block diagram. If no causality can be assigned to a bond graph, no block diagram can be drawn, and so no simulation can be executed with simulation programs, based on the solution of ordinary differential equations.

Remark Note that the causality problem can be solved by putting together two parallel capacities as one capacity, or two inductances in series as one inductance. Moreover, by lumping together two parallel capacitors, the order of the model is reduced and an algebraic dependency between two state variables u_1 and u_2 is removed. Instead of a set of equations that describe a model with differential algebraic equations (DAE with index 2)

$$\dot{u}_1 = i_1/C_1 \tag{2.47}$$

$$\dot{u}_2 = i_2/C_2 \tag{2.48}$$

$$u_1 = u_2 \tag{2.49}$$

a reduced set of equations is obtained which describe a model with ordinary differential equations (ODE)

$$\dot{u} = i/(C_1 + C_2) \tag{2.50}$$

∎

2 – Mass-spring system

To illustrate the use of bond graphs in modeling physical processes, a mass-spring system will be described. This system consists of a mass m [kg], connected with a spring (spring constant k [N/rad]) and a damper (friction constant f [Ns/rad]) to an infinitely large body. The mass is excited by a force F [N]. Derive a model that describes the velocity v [m/s] and position x [m] of the mass.

First of all we assume idealized components. Moreover, we can conclude that there are three forces acting on the mass, namely one exogenous force F and the forces owing to the spring F_s and damper F_f. The resulting force F_m will accelerate the mass.

In using bond graphs we are forced to distinguish the effort and flow variables. Because the proposed process represents a translation of a mechanical system, the force can be selected as effort variable and the velocity as associated flow. Consequently, we have to derive the relations of the four components, namely source F, mass m, spring and damper expressed in force and velocity.

The force F can be represented adequately by means of an effort source S_e. The spring is described by means of

$$F_s = k.x \tag{2.51}$$

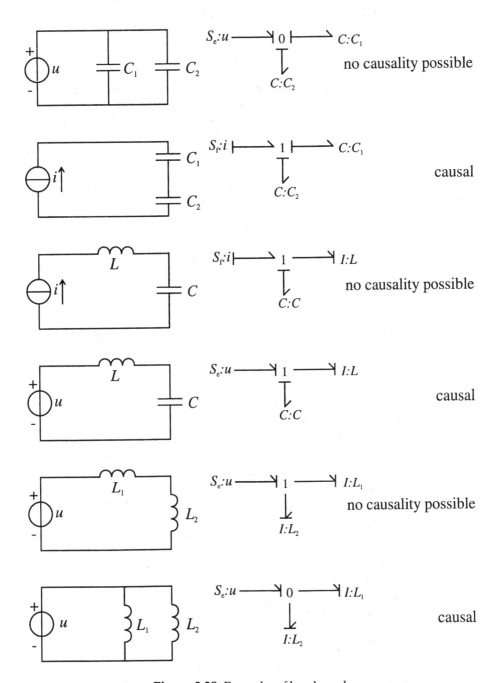

Figure 2.28 Examples of bond graphs

with x the displacement. In fact the application of bond graphs requires us to use efforts and flows. In mechanical translation systems these flows are velocities. Replacing the displacement x by the velocity v, the algebraic equation becomes, with the displacement as the integral of the velocity

$$F_s = k. \int v \, dt \qquad (2.52)$$

Integrating a flow represents the action of a capacity. So, a spring can be represented by means of a capacity with $C = \frac{1}{k}$. The damper represents an element with a friction force F_f which is proportional to the velocity v, or

$$F_f = f.v \qquad (2.53)$$

Clearly, this linear relation between effort F_f and flow v represents a resistor with $R = f$. The resulting force F_m will accelerate the mass m, by

$$F_m = m.a \qquad (2.54)$$

with a the acceleration. Because we have to use the velocity as flow, we have to rewrite this equation as

$$F_m = m\frac{dv}{dt} \qquad (2.55)$$

To avoid differentiators, this equation can be rewritten as an integral equation, namely

$$v = \frac{1}{m} \int F_m \, dt \qquad (2.56)$$

Integrating an effort represents the action of an inertia. So, a mass m can be represented by means of an inertia I with $I = m$.

All four components are connected with each other. This requires a multi-port, either a 0-junction or a 1-junction. It can be observed that in this junction the sum of all efforts is zero and that the velocities are equal for all components, which leads to the selection of a 1-junction. Now we are able to draw the bond graph with the four elements S_e, I (mass), C (spring) and R (damper), as illustrated in Figure 2.29.

In Figure 2.29 the direction of power is defined. Although this choice is arbitrary it is common sense to select this direction from sources to the other components. In the next picture causality has been added. The causality of the source is fixed. Because it is a S_e source, the cross stroke is put at the end. Subsequently, we have two integrators, each with its own causality. The inertia integrates the effort, which indicates that the flow is determined by the effort, shown by a cross stroke at the component side of the bond. The capacity integrates a flow. This implies that the effort is calculated on the basis of a flow, resulting in a cross stroke at the junction side of the bond.

The selection of the causality of the R-element is free. However, the 1-junction requires that all flows are equal. One flow has already been defined, namely by the I-element. Successively, the R-element cannot determine the flow as well. The flow is input for the R-element, which is shown by a cross stroke at the junction side of the bond. Having added the causality and the names of all variables leads to the last bond graph of Figure 2.29.

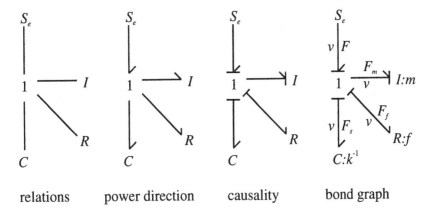

relations   power direction   causality   bond graph

Figure 2.29 Phases in constructing a bond graph

Based on this causal bond graph the associated block diagram can be derived. Block diagrams are centered around the states, being the outputs of the integrators. Starting with them and following the relations of the causal bond graph the other connections can be made. This block diagram is displayed in Figure 2.30.

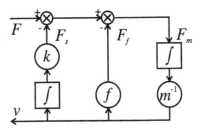

Figure 2.30 Block diagram of mass-spring system

Remark One of the useful characteristics of using the ideas of bond graphs is the necessity of utilizing efforts and flows. Therefore, it was detected in this example that the spring ($F = k.x$) does not represent a resistor R but a capacity C! ∎

Remark This example exhibits only one causal solution. ∎

3 – DC-motor

In a DC-motor a rotor rotates in a stator. The current i through the stator introduces a magnetic field B. If the current i flows through the windings (length l) of the rotor, a force (Lorentz force $F_l = B.i.l$) will be induced on the winding resulting in a torque M which will accelerate the rotor and introduce an angular velocity ω. The basic equations of a DC-motor without losses, as illustrated in Figure 2.31a, are stated as

$$M = m.i \tag{2.57a}$$

$$u = m.\omega \tag{2.57b}$$

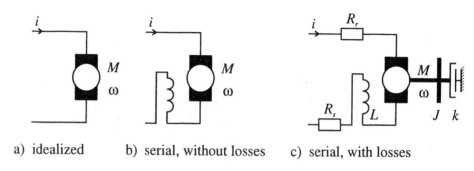

a) idealized   b) serial, without losses   c) serial, with losses

Figure 2.31 Circuits of a DC-motor

According to the selection of effort and flow for both electrical and rotational variables we distinguish the voltage u and torque M as efforts and the rotor current i and angular velocity ω as flows. Then we can recognize in this DC-motor the similarity with a gyrator GY. The effort (M respectively u) on one side influences the flow (i respectively ω) at the other side. The ratio m of the gyrator depends on the stator field. The associated bond graph is shown in Figure 2.32a.

If this stator field is created and maintained by the rotor current (serial use of DC-motor) as illustrated in Figure 2.31b, the ratio m of the gyrator also depends on the current i. The associated bond graph is drawn in Figure 2.32b. Because the DC-motor is still assumed without losses, the bond controlling the value m of the gyrator is an activated bond. The power in the bond before the 1-junction and immediately after this junction has the same value. A more realistic model

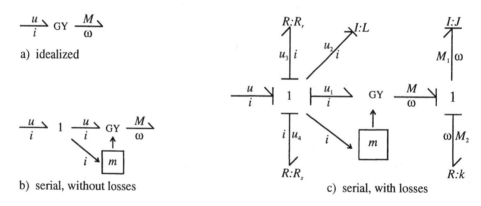

Figure 2.32 Bond graphs of a DC-motor

can be created by introducing losses from resistors both in the stator R_s [Ω] and in the rotor R_r [Ω]. Moreover an inductance L [H] is introduced to represent the inductance in both the stator and the rotor. Furthermore, the motor is assumed to drive a mass m [kg] with inertia J [kgm^2] and experiences a damping k [Ns/rad], as illustrated in Figure 2.31c.

The associated bond graph is drawn in Figure 2.32c. Quite naturally, adding losses introduces new bonds. Friction requires an R-element while inertia and inductance are represented by I-elements. At last, causality is introduced. The causality of the I-elements is fixed. The R and GY elements have two possible causal realizations. Select those realization that fits the rules for the junctions.

2.9 Summary

In this chapter the creation of white-box models of processes is discussed. White-box models are based on prior knowledge and deduction. Black-box models are derived, based on induction, by means of measurements of input and output.

Black-box modeling is discussed in the succeeding chapters. A qualitative model can be derived, based on physical laws. The first step has to be the determination of an appropriate system boundary between the part of the process that will be modeled and the unmodeled part, which is called the environment. A second step is to choose an idealized description of the system with idealized elements. The physical laws that describe the operation of the idealized elements shape the mathematical formulation of the model. It is shown that dynamic models are divided into continuous and discrete models. Continuous models are preferably described by means of ordinary differential equations (ODEs). State events are used to formulate the passing of some threshold value of continuous variables. Discrete models can be formulated by means of difference equations.

If the timing is determined by discrete-event models, difference equations cannot be used. The timing processes of discrete event models are formulated with the aid of processes that shape queues and servers.

It is shown that a set of relations among the variables of a model will naturally result in differential algebraic equations (DAEs). If the modeler is able to assign causality to all variables, the DAE reduces to ordinary differential equations (ODEs). DAEs are characterized by their index. Higher index DAEs are difficult to solve.

Bond graphs provide a general approach for modeling processes from different fields of application. The basic element is a bond. Each bond transports power. The power is the product of an effort across and a flow through the element. In a systematic way the effort and flow can be assigned in each field of application, such as electronics, mechanics and thermal dynamics. Hence, bond graphs assist in selecting the appropriate variables in describing a model.

A second important contribution of bond graphs is the ability to construct first a bond graph without bothering about causality. Later, if all connection are made, well-defined rules allow the assignment of causality to each element.

The chapter concludes with a comparison of circuits, bond graphs and block diagrams. A number a examples are described to illustrate the modeling approach.

2.10 References

Modeling is a basic activity in many scientific areas. Consequently, a vast amount of literature is available. For our purpose we prefer literature that treats modeling as a universal tool, not dedicated to just one application.

A fundamental book is written by Karnopp, Margolis and Rosenberg (1990). They give a concise and thorough treatment of the modeling philosophy and use bond graphs as the unifying theory. Thoma (1975) also introduces bond graphs, but in a much shorter way. Both books concentrate on modeling of physical systems, yielding continuous models.

Other text books dealing with modeling (and partly also with simulation) are written by Cellier (1990) and Kheir (1988). Kheir also discusses models of discrete and discrete event systems.

.11 *Problems*

1. Determine a model of the height $h(t)$ [m] of water in a vertical tank with cross section A [m^3]. The flow out of this tank $F_0(t)$ [m^3/s] is proportional to $h(t)$, so $F_0 = a.h$. The flow into the tank is $F_i(t)$ [m^3/s].

2. In case of the same tank determine also the temperature $T(t)$ [K] of the fluid assuming perfect mixing. Moreover, the specific heat c [J/Kkg] and the specific weight w [kg/m^3] of the fluid are known. The temperature of the incoming flow is $T_i(t)$.
   (Note: The temperature of the flow out of the tank equals the temperature of the fluid into the tank. If there is only a flow out of the tank, the temperature in the tank will not change!).

3. Derive a mathematical model of the following idealized system. In a partly isolated tank there are two flows in and one flow out of the tank. The incoming flows are F_1 [m^3/s] with temperature T_1 [K] and F_2 [m^3/s] with temperature T_2 [K]. A flow F_3 [m^3/s] leaves the tank. The fluid is defined with a specific heat c [J/K.kg] and a specific mass w [kg/m^3]. Owing to the partly isolation of the tank to the environment, there is a energy flow between tank and environment. This flow is determined by the heat resistance k [W/K] and the temperature T_o [K] of the environment. The energy Q [J] stored in the fluid with volume V [m^3] amounts $Q = c.w.V.T$. Assume that the tank is perfectly stirred.

   (a) Define the volume $V(t)$ of the fluid as a function of the flows F_1, F_2 and F_3.

   (b) Define the temperature of the fluid $T(t)$ [K] as a differential equation depending on the physical properties of the tank.

   (c) Show, by an analysis of the physical dimensions, that the resulting expressions for $V(t)$ and $T(t)$ are not incorrect.

   (d) Determine whether F and T are *efforts* or *flows*.

4. A voltage source V [V] with internal resistance R_i [Ω] is connected, by a DC-motor. This motor drives a fan. Motor and fan together have an inertia J [kgm^2]. We assume, for simplicity, that a fan with speed ω [rad/s] requires a torque $M = f.\omega$ [Nm] with f [Nms/rad].

   (a) Draw an acausal bond graph of this system. Indicate with each bond the effort and the flow.

   (b) Add causality. Are there 0, 1,2 or more causal bond graphs? If so, draw all causal bond graphs.

   (c) Draw, if possible, a block diagram of this system.

   (d) If we add inductivity in the wires from voltage source to motor, so L [H], is it still possible to draw a causal bond graph and a block diagram?

5. Determine at least four characteristic differences between a block diagram, a bond graph and a circuit.

6. Determine with the aid of bond graphs whether it is possible to draw a block diagram of two inductances, connected to either a voltage or a current source if these two inductances are connected parallel or in series.

7. What is the difference between a sampled-data model, a time-event model and a state-event model?

8. Derive a bond graph and, if possible, a block diagram of the electronic circuit in Figure 2.33.

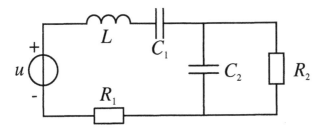

Figure 2.33 Electronic circuit

How many causal models exist?

chapter three

Black-box model representations

3.1 *Introduction*

In Chapter 2 the derivation of models from physical knowledge was described. By means of physical laws (Newton, conservation of mass, Kirchhoff's laws) and the assumption of ideal components, mathematical models can be formulated to describe the behavior of a system. This type of modeling assumes complete knowledge of the process, and is therefore also called *white-box modeling* (see also Figure 2.2).

Although this is a very useful approach, it is not realistic to assume complete knowledge of a process. Depending on the amount of information one has, several situations can occur:

1. The process is too complex to be described by a few idealized physical laws, or there is not enough knowledge available. In this case a general model structure can be used, in which the parameters must all be estimated from measurements of inputs and outputs, using an estimation procedure. This is called *black-box modeling*, reflecting the fact that no knowledge of the process is used.

2. It is possible that the physical laws can be applied to arrive at a model, but that not all parameters in this model are known. For example one does not know the inertia or the friction constants of a system, or the exact value of a capacity is unknown. In this case the physical model is used, but the parameters must be obtained in another way. This case, which is a combination of white-box and black-box modeling, is called *gray-box modeling*. All combinations of black-box modeling and some prior knowledge are called gray-box modeling.

This and the following chapters deal with black-box modeling. Since no information about the system is used, the model must be built from input/output data alone. Contrary to what is discussed in Chapter 2, there are no physical laws on which the model is based. The black-box modeling procedure can be represented as shown in Figure 3.1.

As a first step, as in Chapter 2, the system boundaries must be determined. What will be taken into account as the system, and what will be neglected, and is consequently assumed to be part of the environment? Experiment design is a second step. It is an important step, in which it is decided which signals to measure and what sample frequency to choose, and in which the input signals

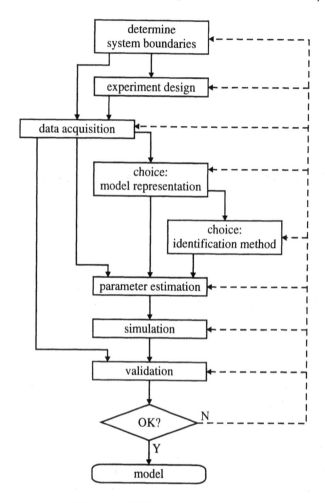

Figure 3.1 Black-box modeling

are selected. Experiment design is postponed until Chapter 6. The next step is data acquisition, the actual collection of the data, needed for identification.

Subsequently, the model representation must be chosen. As will become clear later in this chapter, different model representations can be chosen to describe a system. The choice of the representation depends on the application of the model. Several different black-box model representations are presented in this chapter.

Depending on the model representation chosen, an identification method is selected. This method will yield the estimated parameter values of the model. Several identification methods will be treated in Chapters 4 and 5.

Through simulation the model can be validated with respect to new data. In the validation step the model is evaluated, to see if it can describe the process behavior accurately enough. Model validation is treated in Chapter 6, simulation in Chapter 7.

Before discussing possible model representations, some attention will be paid to naming conventions.

We distinguish between *signals* and *systems*. In our definition a signal is something that carries information, e.g., a current. Signals can be both measurable and unmeasurable.

A system is a set of relations between variables and signals. It is an idealization of the real world. The real world is denoted by *process* or *plant*. In determining a model of a plant, some assumptions are first made about the plant, determining the system. These assumptions can for example concern the boundaries of the process and the selection of inputs, outputs and disturbances.

The signals that we consider are *discrete-time* signals. These are a sequence of numbers, e.g., $\{u(k), k = 1, \ldots, N\}$, that can result from sampling continuous-time signals. One can distinguish *deterministic* and *stochastic* signals. The value of a deterministic signal is fully determined by, e.g., mathematical expressions, tables, etc. Future values of deterministic signals can be predicted, if the signal characteristics and the past values of the signals are known. Stochastic signals are sequences of random numbers. The actual values of a stochastic signal cannot be predicted. The values are described in a statistical framework using mean and variance.

We consider the class of so-called *black-box* models. So, we assume no prior (physical) knowledge on the system. As a consequence, there is no direct relation between the parameters in the model, and the physical properties of the system.

The models under consideration are *discrete-time, linear, time-invariant, causal* models.

The discrete-time notion reflects the fact that the input and output signals are discrete-time signals. The model handles the incoming data (inputs) and produces the outcoming data (outputs) at discrete time instants. Examples of these systems can be found typically in economics and in computer science. But mostly the signals are sampled continuous-time signals: the process itself is continuous-time, but the input and output signals are sampled (sampled-data systems, Section 2.3).

A model is linear if a linear combination of input signals produces as output the same linear combination of the output responses to the individual input signals.

The notion time-invariance should be interpreted as follows: consider an input u, resulting in a corresponding output y. If a time-shifted version of u results in a time-shifted output y, it means that the model's behavior does not depend on time, and the model is called time-invariant. This property is sometimes called *shift-invariance*, for obvious reasons.

Finally, a model is called *nonanticipative, causal* or *proper*, if the output at a certain time instant only depends on the input up to that time instant, including the time instant itself. If the time instant is excluded, the system is called *strictly causal* (or strictly proper).

In this book, *Single-Input Single-Output* (SISO) or *monovariable* systems are considered. This means that there is only one input and one output. The *multivariable* or *Multi-Input Multi-Output* (MIMO) case is not discussed.

A number of different model representations are considered. Both parametric models (finite number of parameters) and nonparametric models (infinite or large number of parameters) will be presented. Also, both time-domain and frequency-domain models are treated.

First something will be said about discrete-time systems, and the consequences of sampling continuous signals. Then in Section 3.3 the well-known

impulse response and transfer function are discussed. In Section 3.4 a characterization of disturbances is provided, using statistical and stochastic properties.
In Sections 3.5 and 3.6 the time-domain and frequency-domain representations
are treated, respectively. Finally, some examples and problems are provided.

3.2 *Discrete-time systems*

In this book discrete-time signals are considered. Usually these signals result
from sampling continuous-time systems (sampled-data systems). A signal is
only measured at discrete, equally spaced, time instants. The time difference
between two time instants is called the sampling time, denoted by T_s.

The continuous signal $u(t)$, $t \in \mathbb{R}$, is represented as the sequence $\{u(k.T_s)\}$,
$k \in \mathbb{Z}$. For simplicity we use a normalized sampling time: $T_s = 1$, although the
theory also applies for other sampling times, after rescaling the time axis. As a
consequence, the signals can be denoted as $\{u(k)\}$, $k \in \mathbb{Z}$.

The sampling frequency ω_s and the sampling time T_s are related according
to

$$\omega_s = \frac{2\pi}{T_s}$$

and since we work with a normalized sampling time

$$\omega_s = 2\pi$$

This means that for a harmonic signal with period ω_0, the following occurs.
Consider the signal

$$u(k) = \cos(\omega_0.k)$$

Then we have, for any $n \in \mathbb{Z}$

$$\cos\left[(\omega_0 + n.\omega_s)k\right] = \cos(\omega_0.k + n.k.2\pi) = \cos(\omega_0.k) = u(k), \quad \forall k, \forall n \in \mathbb{Z}$$

Hence only frequencies in the region $[-\frac{1}{2}\omega_s, \frac{1}{2}\omega_s]$ can be distinguished. Frequencies outside this region are "folded back". This effect is called *aliasing*. To
avoid the problem of aliasing, the sampling frequency must be chosen such that
(Shannon's theorem)

$$\omega_s \geq 2 \cdot \omega_{\max} \tag{3.1}$$

where $\omega_{\max}$ is the maximum frequency present in the signal.

It is assumed that (3.1) is always satisfied. Consequently, for all frequencies ω
we have

$$-\pi \leq \omega \leq \pi$$

Note that the frequency axis is not continuous for a finite number of time
samples N: not all frequencies are present. The frequencies all satisfy

$$\omega = \frac{2\pi r}{N.T_s}, \qquad r \in \left\{-\frac{N}{2}, \ldots, \frac{N}{2}\right\} \tag{3.2}$$

This means that if the number of samples $N \to \infty$, the frequency axis becomes
a continuum between $-\pi$ and π.

Summarizing, we assume that the signals are measured with a normalized
sampling time. The sampling time is chosen such that Shannon's theorem is
obeyed.

All frequencies that are applied lie in the (normalized) interval $[-\pi, \pi]$, and
the frequencies are equally spaced according to (3.2).

3.3 Impulse response and transfer function

Probably the best-known model representations are the *impulse response* and the *transfer function*. The impulse response is the time response of the model, when an impulse is applied to the input. The impulse response is an infinite sequence of numbers, and is given by $g(k)$. For causal systems $g(k) = 0$ for $k < 0$.

The output $y(k)$ of a linear time-invariant and causal model is the convolution of the input $u(k)$ and the impulse response $g(k)$ of the model

$$y(k) = \sum_{\ell=0}^{\infty} g(\ell)u(k - \ell) \tag{3.3}$$

Remark Observe the causality assumption in (3.3): Since the lower bound is $\ell = 0$, the output at time instant k only depends on the input up to and including k.

For *strictly* causal models, the lower bound becomes $\ell = 1$. The output at time instant k then only depends on the input up to (but excluding) k. ∎

The model is completely characterized by its impulse response. This is visualized in Figure 3.2.

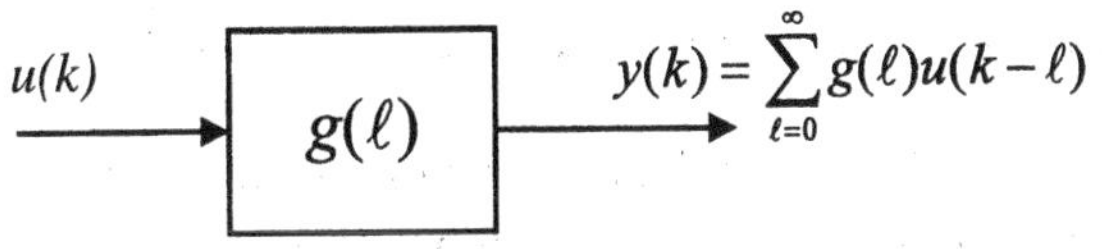

Figure 3.2 Characterization of a discrete-time, linear, time-invariant, causal model by its impulse response

Defining the *forward shift operator* q by

$$qu(k) = u(k + 1)$$

and the *backward shift operator* q^{-1} by

$$q^{-1}u(k) = u(k - 1)$$

we can rewrite (3.3) as

$$y(k) = \sum_{\ell=0}^{\infty} g(\ell) \left[q^{-\ell}u(k) \right] = \left[\sum_{\ell=0}^{\infty} g(\ell)q^{-\ell} \right] u(k) \tag{3.4}$$

The *transfer function* $G(q)$ is now formulated as

$$G(q) = \sum_{\ell=0}^{\infty} g(\ell)q^{-\ell} \tag{3.5}$$

and hence (3.4) can be rewritten as:

$$y(k) = G(q)u(k) \tag{3.6}$$

This is visualized in Figure 3.3.

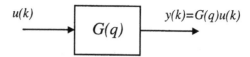

Figure 3.3 Characterization of a discrete-time, linear, time-invariant, causal model by its transfer function

Remark In other contexts, the transfer function is defined as the Z-transform of the sequence $g(\ell)$ for $\ell = 0, 1, 2, \ldots, \infty$, i.e.,

$$G(z) = \sum_{\ell=0}^{\infty} g(\ell) z^{-\ell} \qquad (3.7)$$

The Z-transform is only defined for linear, time-invariant systems. The variable z is a complex number.

As we are working with difference equations we will mainly use the shift operator q, instead of the complex variable z. This implies a slight abuse of notation, but it simplifies matters. Hence our definition (3.6) of the transfer function. ∎

The notion of *stability* is important. We will only use Bounded-Input Bounded-Output (BIBO) stability, which means that if the input is bounded, so is the output.

To investigate the BIBO-stability of a system, we make use of the Z-transform description (3.7), in which z is a complex variable. The *zeros* of the transfer function are the values β_i for which $G(\beta_i) = 0$, and the *poles* of the transfer function are the values α_i for which $G(\alpha_i) = \infty$. In particular, if the transfer function is a rational function of the complex variable z, the zeros are the roots of the numerator polynomial, and the poles are the roots of the denominator polynomial.

With respect to BIBO-stability, the following statements are equivalent:

1. The transfer function $G(q)$ is BIBO-stable

2. $\displaystyle\sum_{\ell=0}^{\infty} |g(\ell)| < \infty$

3. $\displaystyle\lim_{\ell\to\infty} g(\ell) = 0$ and $|g(\ell)|_\infty < \infty$

4. $G(z)$ has all poles inside the unit circle: $|\alpha_i| < 1$.

The transfer function $G(q)$ is *monic* if its first element equals 1, for example

$$G(q) = \sum_{\ell=0}^{\infty} g(\ell) q^{-\ell} \quad \text{with} \quad g(0) = 1$$

Hence we have discussed two nonparametric time-domain model representations: the impulse response (also called Infinite Impulse Response, IIR), and the transfer function. These define a relation between an input signal and an output signal. If the input signal is deterministic, so is the output signal. As will be explained in the next section, this model, as visualized in Figures 3.2 and 3.3, is not very realistic, and it will be extended to include disturbances.

3.4 *Disturbances*

In Section 3.3 the impulse response and the transfer function have been introduced as (nonparametric) model representations in the time-domain. The model, as depicted in Figure 3.3, is not a very practical model to describe a real system. It is assumed that the input is a deterministic signal, and hence fully known. In practice, however, there are always disturbances, usually stochastic, that are beyond our control, and that influence the system as well. These signals can be divided mainly into two classes:

Measurement noise: the sensors, with which the signals are measured, are never ideal. There will always be some error in the so-called *data acquisition*, due to noise and drift.

Uncontrollable inputs: the system can be influenced by signals that can be regarded as inputs, but are not controllable. For instance, one can think of a ship, where the output is its heading. The controllable input is the rudder angle, but the heading will be influenced by the water current as well. This current is an uncontrollable input, and in general unpredictable.

To incorporate the disturbances, they are lumped to the output. This is depicted in Figure 3.4, where all disturbances are lumped in one signal $v(k)$, additive to the output.

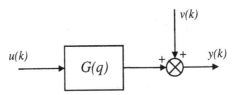

Figure 3.4 System with disturbances lumped to the output

In most cases, the assumption that the disturbances are additive to the system output is valid. Even if, for instance, the measurements of the input are corrupted with noise, in which case the disturbance enters at the input, the error is propagated to the output, and can be added to the signal $v(k)$. In this way also input disturbances can be assumed to be additive to the output.

To describe disturbances, a statistical framework is introduced.

3.4.1 *Statistics and stochastics*

In this section we introduce a statistical and stochastic characterization of disturbances. A disturbance is unpredictable: a value at a future time instant is random, and therefore not known beforehand, even if the past of the signal is known.

Still, disturbances, also known as *stochastic processes* or *random processes*, can be characterized by their statistical and stochastic properties.

The distinction between statistics and stochastics might not be clear at first sight. The *statistical* properties of a signal describe the amplitude distribution of the signal: how often does a certain amplitude occur? It does not matter in what order the values of the signal occur. The *stochastic* properties of a signal describe the distribution of frequencies in the signal: does there exist a dependency between the amplitude at this moment, and the amplitude at the next time instant? Several definitions of statistical and stochastic properties are given, after which the difference between the two is illustrated in Example 3.1.

By definition a stochastic process consists of a number of realizations. In any real situation only one single realization is available. For example, if a signal is measured, the measurement noise is one realization of a stochastic process. Therefore we want to use stochastic processes that are completely determined by the stochastic properties of one single realization (also called a *time series*). Hence, to describe these processes, one realization must be sufficient. Moreover, we do not want the characteristics to depend on the particular time instant at which they are measured. Stochastic processes with these two properties are called *weakly stationary, weakly ergodic* processes. A more formal definition of these properties is postponed to a later point in this section.

From now on we assume that all disturbance signals are a realization of a weakly ergodic, weakly stationary stochastic process, unless stated otherwise. These signals will be called *disturbance signals* or *stochastic signals*.

We will now turn to the characterization of stochastic signals. The first definition is concerned with both a statistical and a stochastic property.

Definition 3.1 (Mean value)
For the discrete-time stochastic signal $v(k)$, the mean value *or first statistical moment is defined as:*

$$\bar{v} = \mathbb{E}\{v(k)\} = \lim_{N \to \infty} \frac{1}{N} \sum_{k=1}^{N} v(k) \tag{3.8}$$

where $\mathbb{E}$ is the expectation operator.

The expectation operator $\mathbb{E}$ is a linear operator. Hence for any two stochastic signals $v(k)$ and $w(k)$, and for any two constants c_1 and c_2, the following holds

$$\mathbb{E}\{c_1 v(k) + c_2 w(k)\} = c_1 \mathbb{E}\{v(k)\} + c_2 \mathbb{E}\{w(k)\} = c_1 \bar{v} + c_2 \bar{w} \tag{3.9}$$

A second statistical property of stochastic signals is the *variance*:

Definition 3.2 (Variance)
For the discrete-time stochastic signal $v(k)$, the variance *or second statistical moment is defined as*

$$\sigma_v{}^2 = \mathbb{E}\{[v(k) - \bar{v}]^2\} \tag{3.10}$$

Now the stochastic properties of a stochastic signal are described. The (auto)covariance function is defined as follows.

Definition 3.3 (Autocovariance function)
For the discrete-time stochastic signal $v(k)$, the (auto)covariance function *is defined as*

$$R_v(\tau) = \mathbb{E}\{[v(k) - \bar{v}][v(k - \tau) - \bar{v}]\} \qquad \tau \in \mathbb{Z} \tag{3.11}$$

Remark The covariance function is symmetric: $R_v(\tau) = R_v(-\tau)$.   ∎

Remark There is a clear relationship between the variance (statistical property) and the covariance (stochastic property) of a signal

$$\sigma_v{}^2 = R_v(0) \tag{3.12}$$

That is, the variance equals the covariance for lag 0.   ∎

The (auto)correlation function is defined as:

Definition 3.4 (Correlation function)
For the discrete-time stochastic signal $v(k)$, the (auto)correlation function *is defined as follows*

$$X_v(\tau) = \mathbb{E}\{v(k)v(k-\tau)\} \qquad \tau \in \mathbb{Z} \tag{3.13}$$

Remark The correlation function is symmetric: $X_v(\tau) = X_v(-\tau)$. ∎

Remark In some textbooks the property defined by (3.13) is called the *cross product*, and the term autocorrelation is reserved for the expression $\mathbb{E}\{v(k)v(k-\tau)\}/\mathbb{E}\{v(k)v(k)\}$, which has values between -1 and 1. By sticking to the definition (3.13), we do not restrict ourselves: if a *scaled* signal is used (that is, all signal values are divided by the variation: square root of the variance) we allow the same quantity to be denoted by the correlation. ∎

Remark The relation between the covariance function and the correlation function of a signal v is

$$R_v(\tau) = X_v(\tau) - \bar{v}^2 \tag{3.14}$$

It is then clear that the covariance and the correlation functions are equal if the signal v has zero mean. ∎

Similarly, the cross-covariance function for two discrete-time stochastic signals $v(k)$ and $w(k)$ is defined by

$$R_{vw}(\tau) = \mathbb{E}\{[v(k)-\bar{v}][w(k-\tau)-\bar{w}]\} \qquad \tau \in \mathbb{Z} \tag{3.15}$$

and the cross-correlation function is given by

$$X_{vw}(\tau) = \mathbb{E}\{v(k)w(k-\tau)\} \qquad \tau \in \mathbb{Z} \tag{3.16}$$

Remark The signals v and w in the definitions of the cross-covariance function (3.15) and the cross-correlation function (3.16) can be interchanged according to

$$R_{vw}(\tau) = R_{wv}(-\tau) \tag{3.17a}$$
$$X_{vw}(\tau) = X_{wv}(-\tau) \tag{3.17b}$$

∎

Remark There exists a relation between the cross-covariance function and the cross-correlation function:

$$R_{vw}(\tau) = X_{vw}(\tau) - \bar{v} \cdot \bar{w} \tag{3.18}$$

Again, if one of the signals v or w has zero mean, the cross-covariance and the cross-correlation functions are equal. ∎

The terms *ergodicity* and *stationarity* have already been used. Now that several stochastic properties have been defined, it is possible to give a formal definition of weak stationarity and weak ergodicity.

Definition 3.5 (Weak stationarity)
A stochastic signal $v(k)$ is weakly stationary if the following three conditions are satisfied:

1. The signal power has a finite upper bound: $\mathbb{E}\{v^2(k)\} < \infty$

2. The mean value is time-invariant: $\mathbb{E}\{v(k)\} = c$

3. The autocorrelation function only depends on the time difference τ, and not on the particular time instants k at which it is calculated.

Definition 3.6 (Weak ergodicity)
A stochastic signal $v(k)$ is weakly ergodic if the following three conditions are satisfied:

1. The signal $v(k)$ is weakly stationary

2. The expectation (ensemble average) is equal to the mean value (time average) (cf. (3.8)).

3. The ensemble autocorrelation function is equal to the time autocorrelation function

One more definition will be used, namely the *covariance matrix*.

Definition 3.7 (Covariance matrix)
The covariance matrix $\mathcal{R}_v^n$ *of the stochastic signal* $v(k)$ *is defined as the expectation of the outer product of the vector of n past values of $v(k)$ with itself.*

$$\mathcal{R}_v^n = \mathbb{E}\left\{ \begin{pmatrix} v(k) - \bar{v} \\ v(k-1) - \bar{v} \\ \vdots \\ v(k-n+1) - \bar{v} \end{pmatrix} \times \begin{pmatrix} v(k) - \bar{v} \\ v(k-1) - \bar{v} \\ \vdots \\ v(k-n+1) - \bar{v} \end{pmatrix} \right\} \tag{3.19}$$

The covariance matrix $\mathcal{R}_v^n$ can be written as

$$\mathcal{R}_v^n = \begin{bmatrix} R_v(0) & R_v(1) & \cdots & R_v(n-1) \\ R_v(-1) & R_v(0) & \cdots & R_v(n-2) \\ \vdots & \vdots & \ddots & \vdots \\ R_v(1-n) & R_v(2-n) & \cdots & R_v(0) \end{bmatrix} \tag{3.20}$$

Remark Since $\mathbb{E}\{[v(k)-\bar{v}][v(k-1)-\bar{v}]\} = \mathbb{E}\{[v(k-1)-\bar{v}][v(k)-\bar{v}]\}$ for weakly ergodic processes, the covariance matrix is symmetric. ∎

For weakly ergodic processes, the expectation operator is given by (3.8). This means that it can be calculated if the number of measurements, N, tends to infinity. This is not a very practical situation. Therefore in practice the expectation must be estimated on the basis of a finite number of measurements

$$\hat{\bar{v}} = \frac{1}{N} \sum_{k=1}^{N} v(k) \tag{3.21}$$

$$\hat{R}_v(\tau) = \frac{1}{N} \sum_{k=1}^{N} [v(k) - \hat{\bar{v}}][v(k - \tau) - \hat{\bar{v}}] \tag{3.22}$$

where $\hat{\cdot}$ denotes that it is an estimate. In Section 3.7 an example is given to indicate the accuracy of this estimate.

Now that we have given some definitions of statistical and stochastic properties, the difference between them can be shown by the following example.

Example 3.1 Consider the signals that are shown in Figure 3.5. Note that they are not stochastic processes, but the theory still applies. Deterministic signals are chosen to better illustrate the difference between statistical and stochastic properties.

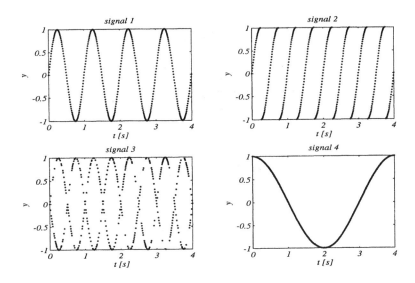

Figure 3.5 Four signals with the same mean and variance

The different signals are

1. $y(k) = \sin(\omega_0 k)$

2. $y(k) = \sin(\omega_0 k).\text{sign}[\cos(\omega_0 k)]$

3. $y(k) = \sin(\omega_0 k).\text{sign}[\sin(50\omega_0 k)]$

4. $y(k) = \cos(\frac{1}{4}\omega_0 k)$

They all have the same statistical properties: mean ($= 0$) and variance are equal for all four signals. The only difference between the signals is the ordering of the amplitude.

The stochastic properties, however, are quite different. The covariance functions are:

1. $R_y(\tau) = \frac{1}{2}\cos(\omega_0 \tau)$

2. $R_y(\tau) = \frac{1}{2}\cos(\omega_0 \tau).\text{tri}_C(\omega_0 \tau)$

3. $R_y(\tau) = \frac{1}{2}\cos(\omega_0 \tau).\text{tri}_C(50\omega_0 \tau)$

4. $R_y(\tau) = \frac{1}{2}\cos(\frac{1}{4}\omega_0 \tau)$

The function $\text{tri}_C(\omega_0 \tau)$ is a periodic triangular function, plotted in Figure 3.6. The phase equals the phase of $\cos(\omega_0 t)$.

In Figure 3.7 the covariance functions are shown for all four signals.

This demonstrates why statistical properties are not sufficient, and why in addition stochastic properties should be used in describing the dynamic behavior of stochastic signals.   □

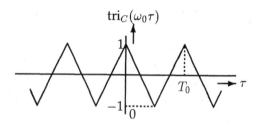

Figure 3.6 The function $\mathrm{tri}_C(\omega_0\tau)$

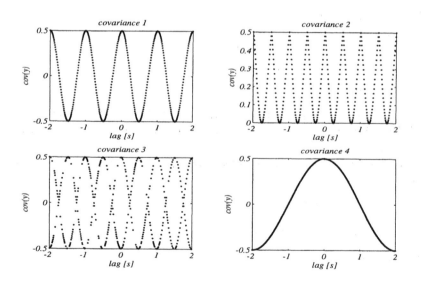

Figure 3.7 Covariance functions of the signals from Figure 3.5

In the next subsection a strictly stationary ergodic stochastic process will be introduced, with which all disturbances are described.

3.4.2 Discrete white noise

Almost all stochastic processes of interest can be assumed to be generated by a special random process, applied to a linear filter. This special random process has some nice stochastic characteristics: it generates "shocks", that are not correlated. It is called *white noise*, and it is denoted by $e(k)$. A white noise process is described entirely by its first and second statistical moments. A special case of white noise sequences is the *Zero Mean White Noise* (ZMWN) sequence, which is defined by

$$\mathbb{E}\{e(k)\} = 0 \quad \text{for all } k \tag{3.23a}$$

$$\mathbb{E}\{e(k)e(\ell)\} = \begin{cases} \sigma_e^2 & \text{for } k = \ell \\ 0 & \text{otherwise} \end{cases} \tag{3.23b}$$

Remark The whiteness of a stochastic process has to deal with stochastic properties: the distribution of the frequencies. All frequencies are equally important just as in white light. Nothing is said about the statistical properties: the distribution of the amplitude. For white noise, the value $e(k)$ at instant k can have a normal (or Gaussian) distribution, but it can equally well be uniformly distributed, or have any other distribution. In practice we usually assume that the noise at instant k has a normal distribution, denoted by $\mathcal{N}(0, \sigma_e^2)$. ∎

As was said before, the disturbances we consider are generated by a white noise sequence, which is filtered by a linear filter. Hence, all stochastic signals $v(k)$ can be written as

$$v(k) = \sum_{\ell=0}^{\infty} h(\ell)e(k - \ell)$$

or, equivalently:

$$v(k) = H(q)e(k) \tag{3.24}$$

The transfer function $H(q)$ is assumed to be monic, and hence $h(0) = 1$. Since $v(k)$ is filtered white noise, it is also called *colored noise* (Note again the similarity with colored light!).

Now that we have a characterization of disturbances, through the definition of white noise, the disturbance v in our model in Figure 3.4 can be described more explicitly. This is shown in Figure 3.8.

Figure 3.8 shows a model, incorporating both a deterministic part, $G(q)u(k)$, and a stochastic part, $H(q)e(k)$. The output $y(k)$ becomes

$$y(k) = G(q)u(k) + H(q)e(k) \tag{3.25}$$

Model (3.25) represents our general black-box model.

The model representations used until now are nonparametric time-domain descriptions: the impulse response and the transfer function as defined in (3.3) and (3.6), respectively, have an infinite number of parameters. In the next two sections other representations are discussed, in both time and frequency domain.

In the remaining part of this book the following assumptions are made:

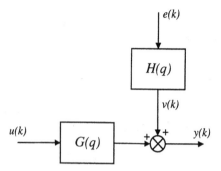

Figure 3.8 Model with disturbance lumped to the output

1. The deterministic transfer function $G(q)$ is strictly proper, which means that $g(\ell) = 0$ for $\ell \leq 0$.

2. The stochastic transfer function $H(q)$ is monic, and proper, which means that $h(\ell) = 0$ for $\ell < 0$, and $h(0) = 1$.

3. The systems under consideration are linear and time-invariant.

4. All stochastic signals are generated by weakly ergodic stochastic processes.

3.5 *Time-domain representations*

Dealing with models, either for simulation, prediction or design, becomes easier if the number of parameters in the model is finite, and preferably small. From that point of view nonparametric models are not attractive, and parametric models are preferred.

In this section five parametric time-domain model representations will be discussed: the Finite Impulse Response (FIR) model, the Auto-Regressive with eXogenous input (ARX) model, the Auto-Regressive Moving Average with eXogenous input (ARMAX) model, the Output Error (OE) model and the Box-Jenkins (BJ) model. These models are special cases of the following general model:

$$A(q)y(k) = \frac{B(q)}{F(q)}u(k) + \frac{C(q)}{D(q)}e(k) \tag{3.26}$$

where $A(q)$, $B(q)$, $C(q)$, $D(q)$ and $F(q)$ are polynomials in the shift operator q, defined as

$$
\begin{aligned}
A(q) &= 1 + a_1 q^{-1} + a_2 q^{-2} + \cdots + a_{n_a} q^{-n_a} \\
B(q) &= \phantom{1 + {}} b_1 q^{-1} + b_2 q^{-2} + \cdots + b_{n_b} q^{-n_b} \\
C(q) &= 1 + c_1 q^{-1} + c_2 q^{-2} + \cdots + c_{n_c} q^{-n_c} \\
D(q) &= 1 + d_1 q^{-1} + d_2 q^{-2} + \cdots + d_{n_d} q^{-n_d} \\
F(q) &= 1 + f_1 q^{-1} + f_2 q^{-2} + \cdots + f_{n_f} q^{-n_f}
\end{aligned}
$$

By making certain choices for these polynomials, the different model representations are obtained.

A more formal term for model representation is *model structure*: the model is structured by the use of predefined polynomials.

Of course it would suffice to use the general model (3.26) in all situations. However, it will prove to be convenient in our calculations to simplify the general model, by fixing some of the polynomials at 1. For example, the FIR- and ARX-structures lead to linear equations in the identification algorithms. This will be made more explicit in Chapter 5.

3.5.1 Finite Impulse Response model

The Infinite Impulse Response (IIR) model (3.3) is given by

$$y(k) = \sum_{\ell=1}^{\infty} g(\ell)u(k-\ell) + e(k)$$

Since $G(q)$ is stable, there exists a $M < \infty$, for which the following holds

$$g(\ell) = 0 \quad \text{for} \quad \ell > M$$

Hence only the first M values of the impulse response have to be taken into account, which results in the following Finite Impulse Response (FIR) model

$$y(k) = \sum_{\ell=1}^{M} g(\ell)u(k-\ell) + e(k)$$

This model can be obtained from the general model (3.26), by choosing $A(q) = C(q) = D(q) = F(q) = 1$, and $B(q)$ an arbitrary polynomial of order M. This gives the standard FIR model

$$y(k) = B(q)u(k) + e(k) \tag{3.27}$$

Note that the disturbances are modeled as white noise, so $H(q) = 1$.

3.5.2 Auto-Regressive with eXogenous input model

The Auto-Regressive with eXogenous input (ARX) model can be obtained from the general model (3.26) by choosing $C(q) = D(q) = F(q) = 1$, and $A(q)$ and $B(q)$ arbitrary polynomials

$$A(q)y(k) = B(q)u(k) + e(k) \tag{3.28}$$

Since the noise enters directly in the equation, the model is of the class of *equation error* models. The equation error is modeled as a white noise sequence. The AR-part comes from the fact that the output is a function of its past values (A-polynomial). The input u is the exogenous (or extra) input.
The model (3.28) can also be written as

$$y(k) = \frac{B(q)}{A(q)}u(k) + \frac{1}{A(q)}e(k) \tag{3.29}$$

to reflect the transfer functions $G(q)$ and $H(q)$.
Note that the disturbances are modeled as AR-filtered white noise, additive to the output.

3.5.3   *Auto-Regressive Moving Average with eXogenous input model*

The Auto-Regressive Moving Average with eXogenous input (ARMAX) model
can be derived from the general model (3.26) by choosing $D(q) = F(q) = 1$, and
$A(q)$, $B(q)$ and $C(q)$ arbitrary polynomials

$$A(q)y(k) = B(q)u(k) + C(q)e(k) \tag{3.30}$$

or, alternatively

$$y(k) = \frac{B(q)}{A(q)}u(k) + \frac{C(q)}{A(q)}e(k) \tag{3.31}$$

The ARMAX model is of the class of equation error models. The equation error
is modeled as a MA-process. The disturbances are modeled as white noise,
passed through an ARMA filter, and additive to the output.

3.5.4   *Output Error model*

The Output Error (OE) model can be derived from the general model (3.26), by
choosing $A(q) = C(q) = D(q) = 1$, and $B(q)$ and $F(q)$ arbitrary polynomials. In
this case the (white) noise is supposed to be additive to the output. To show this
structure explicitly, an auxiliary, noise-free signal $w(k)$ is usually introduced:

$$F(q)w(k) = B(q)u(k) \tag{3.32a}$$
$$y(k) = w(k) + e(k) \tag{3.32b}$$

This can be written alternatively as

$$y(k) = \frac{B(q)}{F(q)}u(k) + e(k) \tag{3.33}$$

Note that the plant model, $G(q) = B(q)/F(q)$, and the noise model, $H(q) = 1$,
have no parameters in common. Therefore they are said to be "independently
parametrized".

3.5.5   *Box-Jenkins model*

The Box-Jenkins (BJ) model can be derived from (3.26) by choosing $A(q) = 1$,
and the other polynomials arbitrarily

$$y(k) = \frac{B(q)}{F(q)}u(k) + \frac{C(q)}{D(q)}e(k) \tag{3.34}$$

The disturbances are modeled as ARMA-filtered white noise, additive to the
output.
Note that, as in the OE-case, the plant model, $G(q) = B(q)/F(q)$, and the noise
model, $H(q) = C(q)/D(q)$, have no parameters in common. Hence they are
independently parametrized.

3.5.6 Summary

To summarize the models described in this section, Table 3.1 is made. Starting from the general model (3.26), it shows how the polynomials should be chosen to obtain a certain model structure. A '1' denotes that the polynomial must be fixed to 1, a '×' denotes that it can be chosen freely.

Table 3.1
Different model structures and the choice of the polynomials

	A	B	C	D	F	Equation
General	×	×	×	×	×	(3.26)
FIR	1	×	1	1	1	(3.27)
ARX	×	×	1	1	1	(3.28)
ARMAX	×	×	×	1	1	(3.30)
OE	1	×	1	1	×	(3.32)
BJ	1	×	×	×	×	(3.34)

The main difference between the model structures is the way the disturbances are modeled.

The ARX and ARMAX model structures are of the equation error type, the OE and BJ model structures are of the output error type. The latter type of models has independently parametrized plant and noise models, whereas in the equation error type of models the plant and noise model have parameters in common.

The following example illustrates the different model structures.

Example 3.2 The different black-box model structures can be written as difference equations as follows.

FIR:
$$y(k) = b_1 u(k-1) + b_2 u(k-2) + e(k)$$

ARX:
$$y(k) = b_1 u(k-1) + b_2 u(k-2) - a_1 y(k-1) - a_2 y(k-2) + e(k)$$

ARMAX:
$$y(k) = b_1 u(k-1) + b_2 u(k-2) - a_1 y(k-1) - a_2 y(k-2) + e(k) + c_1 e(k-1) + c_2 e(k-2)$$

OE:
$$w(k) = b_1 u(k-1) + b_2 u(k-2) - f_1 w(k-1) - f_2 w(k-2)$$
$$y(k) = w(k) + e(k)$$

BJ:
$$w(k) = b_1 u(k-1) + b_2 u(k-2) - f_1 w(k-1) - f_2 w(k-2)$$
$$v(k) = e(k) + c_1 e(k-1) + c_2 e(k-2) - d_1 v(k-1) - d_2 v(k-2)$$
$$y(k) = w(k) + v(k)$$

In the next section several frequency-domain model representations are discussed.

3.6 *Frequency-domain representations*

In the previous section some time-domain model representations were given. As it is sometimes advantageous to describe systems in the frequency domain, a number of frequency-domain model representations will be discussed in this section.

3.6.1 *Frequency function*

The frequency-domain equivalent of the transfer function $G(q)$ in (3.6) is the *frequency function*. It is obtained by replacing q by $e^{i\omega}$. This is clarified as follows.

Suppose that the system (3.6) is excited by a sinusoidal signal

$$u(k) = \cos(\omega_0 k) \qquad -\pi \le \omega_0 \le \pi \qquad (3.35)$$

For convenience (3.35) is written as

$$u(k) = \mathrm{Re}\left\{e^{i\omega_0 k}\right\} \qquad (3.36)$$

where 'Re' denotes the real part of a complex number. The corresponding output becomes with (3.3)

$$
\begin{aligned}
y(k) &= \sum_{\ell=1}^{\infty} g(\ell)\mathrm{Re}\left\{e^{i\omega_0(k-\ell)}\right\} = \mathrm{Re}\left\{\sum_{\ell=1}^{\infty} g(\ell)e^{i\omega_0(k-\ell)}\right\} \\
&= \mathrm{Re}\left\{e^{i\omega_0 k}\sum_{\ell=1}^{\infty} g(\ell)e^{-i\omega_0\ell}\right\} = \mathrm{Re}\left\{e^{i\omega_0 k}G(e^{i\omega_0})\right\} \\
&= \left|G(e^{i\omega_0})\right| \cos(\omega_0 k + \phi) \qquad (3.37\text{a})
\end{aligned}
$$

where

$$\phi = \arg G(e^{i\omega_0}) \qquad (3.37\text{b})$$

Remark In (3.37a) it is assumed that the signal u has been applied to the system for a long period: transient effects are neglected. If this is not the case, a term must be added containing the transient effects. This term decays for all stable systems.   ∎

From (3.37a) it can be seen that the output of the linear system with transfer function $G(q)$, to a sinusoid with frequency ω_0, will be a sinusoid with the same frequency, with its amplitude multiplied by $\left|G(e^{i\omega_0})\right|$, and a phase shift of $\arg G(e^{i\omega_0})$ radians.

The complex number $G(e^{i\omega_0})$ is the transfer function of the system, evaluated at $q = e^{i\omega_0}$. It gives complete information about what happens if the input is a sinusoid with frequency ω_0.

Of course we can evaluate $G(e^{i\omega})$ over the complete unit circle, that is for $-\pi \le \omega \le \pi$. These values for G are called the *frequency function*:

$$G(e^{i\omega}) \qquad \text{for} \qquad -\pi \le \omega \le \pi \qquad (3.38)$$

Usually the frequency function (3.38) is graphically displayed in a *Bode plot* or a *Nyquist plot*. In the Bode plot both the amplitude and the phase are plotted as a function of the frequency ω. In a Nyquist plot the frequency function is plotted in the complex plane.

3.6.2   Signal spectrum

A stationary stochastic process can be described completely by its mean and covariance function. This is a time-domain description. A frequency-domain description of these processes is the (power) spectrum. It is defined as the Fourier transform of the covariance function, and can be interpreted as the information on the frequency contents of a stochastic process. Before formally introducing the signal spectrum, one other thing has to be discussed.

The covariance function was defined for stochastic signals only. To include deterministic signals in our description, we introduce a generalized expectation operator $\bar{\mathbb{E}}$:

$$\bar{\mathbb{E}} = \frac{1}{N} \sum_{k=1}^{N} \mathbb{E} \tag{3.39}$$

Note that for stochastic signals the generalized expectation operator $\bar{\mathbb{E}}$ equals the expectation operator $\mathbb{E}$, since the summation has no influence here. For deterministic signals the stochastic expectation has no effect, and hence the summation applies here. Now all definitions of Section 3.4.1 are valid for deterministic signals, if we replace $\mathbb{E}$ by $\bar{\mathbb{E}}$.

We have assumed all stochastic signals to be stationary and ergodic. A similar assumption is made for deterministic signals. Combining the two, we assume from now on that all signals we use, stochastic or deterministic, are *quasi-stationary* according to the following definition:

Definition 3.8 *A signal, stochastic or deterministic, is called* quasi-stationary *if the following holds:*

1. $\bar{\mathbb{E}}\{s(t)\} = \bar{s}(t)$ *is bounded for all* $t \in \mathbb{Z}$

2. $\bar{\mathbb{E}}\{[s(t) - \bar{s}(t)][s(r) - \bar{s}(r)]\} = R_s(t, r)$ *is bounded for all* $t, r \in \mathbb{Z}$ *and*
   $$\lim_{N \to \infty} \frac{1}{N} \sum_{t=1}^{N} R_s(t, t - \tau) = R_s(\tau), \quad \forall t, \tau \in \mathbb{Z}$$

Hence we assume that the mean and the energy of the signal are bounded.

Since from now on the definitions of covariance and cross covariance will be used for both deterministic and stochastic signals, the following definitions of spectrum and cross spectrum are general definitions.

Definition 3.9 (Power spectrum)
The (power) spectrum $\Phi_v(\omega)$ *of a time series* $\{v(k)\}$ *is defined as the Fourier transform of the covariance function* $R_v(\tau)$ *(3.11) of that time series:*

$$\Phi_v(\omega) = \sum_{\tau=-\infty}^{\infty} R_v(\tau) e^{-i\tau\omega} \tag{3.40}$$

and the cross spectrum *between* $\{v(k)\}$ *and* $\{w(k)\}$ *as*

$$\Phi_{vw}(\omega) = \sum_{\tau=-\infty}^{\infty} R_{vw}(\tau) e^{-i\tau\omega} \tag{3.41}$$

provided the infinite sums exist.

The spectrum $\Phi_v(\omega)$ is a real-valued and symmetric function of ω. The cross spectrum $\Phi_{vw}(\omega)$ is in general a complex-valued function of ω. Its real part is known as the *cospectrum*, and its imaginary part is called the *quadrature spectrum*. It can also be decomposed into its *amplitude spectrum* ($|\Phi_{vw}(\omega)|$) and its *phase spectrum* ($\arg \Phi_{vw}(\omega)$).

Remark Note that, by definition of the Inverse Fourier Transform, we have

$$R_v(0) = \frac{1}{2\pi} \int_{-\pi}^{\pi} \Phi_v(\omega) \, d\omega \qquad N \to \infty \tag{3.42}$$

∎

The spectrum is illustrated by the following two examples.

Example 3.3 (Spectrum of a sinusoid)
Consider the signal

$$u(k) = A \cos(\omega_0 k) \qquad -\pi \le \omega_0 \le \pi \tag{3.43}$$

To calculate the spectrum $\Phi_u(\omega)$, (3.40) is used. The covariance function must then be calculated first. It is given by

$$R_u(\tau) = \lim_{N \to \infty} \frac{1}{N} \sum_{k=1}^{N} u(k) u(k - \tau)$$

$$= \lim_{N \to \infty} \frac{1}{N} \sum_{k=1}^{N} A^2 \cos(\omega_0 k) \cos[\omega_0(k - \tau)] \tag{3.44}$$

From goniometrics it is known that

$$\cos(\omega_0 k) \cos[\omega_0(k - \tau)] = \frac{1}{2} \left[\cos(2\omega_0 k - \omega_0 \tau) + \cos(\omega_0 \tau) \right]$$

Then (3.44) can be rewritten as

$$R_u(\tau) = \lim_{N \to \infty} \frac{1}{N} \sum_{k=1}^{N} \left[\frac{A^2}{2} \cos(2\omega_0 k - \omega_0 \tau) + \frac{A^2}{2} \cos(\omega_0 \tau) \right]$$

$$= \frac{A^2}{2} \cos(\omega_0 \tau)$$

To arrive at the latter equality, use is made of the following fact:

$$\lim_{N \to \infty} \frac{1}{N} \sum_{k=1}^{N} \cos(\omega k + \varphi) = 0 \qquad \forall \omega, \varphi$$

Then the spectrum is

$$\Phi_u(\omega) = \sum_{\tau=-\infty}^{\infty} R_u(\tau) e^{-i\tau\omega}$$

$$= \sum_{\tau=-\infty}^{\infty} \frac{A^2}{2} \cos(\omega_0 \tau) e^{-i\tau\omega}$$

Rewriting $\cos(\omega_0 \tau)$ as

$$\cos(\omega_0 \tau) = \frac{e^{i\omega_0 \tau} + e^{-i\omega_0 \tau}}{2}$$

we obtain

$$\Phi_u(\omega) = \frac{A^2}{4} \sum_{\tau=-\infty}^{\infty} \left[e^{-i(\omega-\omega_0)\tau} + e^{-i(\omega+\omega_0)\tau} \right]$$

$$= \frac{A^2 \pi}{2} \left[\delta(\omega - \omega_0) + \delta(\omega + \omega_0) \right] \qquad -\pi \le \omega \le \pi \qquad (3.45)$$

where $\delta(\omega)$ is Dirac's Delta function, defined as

$$\begin{cases} \int_{-\infty}^{\infty} \delta(\omega)\, d\omega &= 1 \quad , \quad \omega \in \mathbb{R} \\ \delta(\omega) &= 0 \quad , \quad \omega \ne 0 \end{cases}$$

By applying the Inverse DFT to (3.45), we obtain the original $u(k)$, which proves the derivation.

Hence, the spectrum of a sinusoid with frequency ω_0 is nonzero at $\omega = \pm\omega_0$, and zero elsewhere. This is illustrated in Figure 3.9.

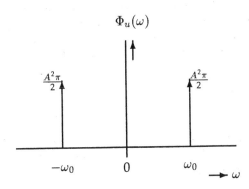

Figure 3.9 Spectrum of a sinusoid with amplitude A

☐

Example 3.4 (Spectrum of discrete white noise)
Consider the Gaussian distributed Zero Mean White Noise sequence $e(k)$. The variance of this signal is σ_e^2. The covariance function is given by

$$R_e(\tau) = \sigma_e^2 \Delta(\tau)$$

where $\Delta(\tau)$ is the unit pulse, defined as

$$\Delta(\tau) = \begin{cases} 1 & \tau = 0 \\ 0 & \tau \ne 0, \ \tau \in \mathbb{Z} \end{cases}$$

The spectrum of the signal e is then computed as

$$\Phi_e(\omega) = \sum_{\tau=-\infty}^{\infty} R_e(\tau) e^{-i\tau\omega} = \sigma_e^2 \qquad -\pi \le \omega \le \pi \qquad (3.46)$$

It is drawn in Figure 3.10. Since $e(k)$ is a discrete signal, its spectrum is restricted to the interval $[-\pi, \pi]$.

☐

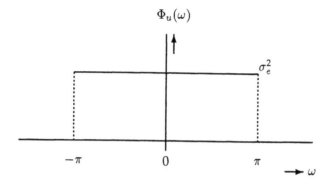

Figure 3.10 Spectrum of discrete white noise

3.6.3   *Transformation of spectra*

In describing systems, we are interested in how a system reacts to a certain input signal: given an input signal, what is the resulting output signal?
Hence we are interested in the relation between spectra of input signals and spectra of output signals. In this section this relation is established.

   Consider the output signal y of the general model, given by (3.25):

$$y(k) = G(q)u(k) + H(q)e(k) = \sum_{\ell=1}^{\infty} g(\ell)u(k-\ell) + \sum_{\ell=0}^{\infty} h(\ell)e(k-\ell) \quad (3.47)$$

To compute the spectrum of y, we first need its covariance function. Consider the case where $\bar{u} = \bar{e} = 0$, and where u and e are uncorrelated: $R_{ue}(\tau) = 0 \; \forall \tau \in \mathbb{Z}$. The covariance function of y is given by

$$
\begin{aligned}
R_y(\tau) = \mathbb{E}\{y(k)y(k-\tau)\} = \mathbb{E} &\left\{ \left[\sum_{\ell=1}^{\infty} g(\ell)u(k-\ell) + \sum_{\ell=0}^{\infty} h(\ell)e(k-\ell) \right] \cdot \right. \\
&\left. \left[\sum_{m=1}^{\infty} g(m)u(k-\tau-m) + \sum_{m=0}^{\infty} h(m)e(k-\tau-m) \right] \right\} \\
= \mathbb{E} &\left\{ \sum_{\ell=1}^{\infty} g(\ell)u(k-\ell) \sum_{m=1}^{\infty} g(m)u(k-\tau-m) \right\} \\
+ \mathbb{E} &\left\{ \sum_{\ell=0}^{\infty} h(\ell)e(k-\ell) \sum_{m=0}^{\infty} h(m)e(k-\tau-m) \right\} \quad (3.48)
\end{aligned}
$$

since $\mathbb{E}$ is linear, and since u and e are uncorrelated. The covariance function (3.48) can be rewritten as

$$
\begin{aligned}
R_y(\tau) = \sum_{\ell=1}^{\infty} \sum_{m=1}^{\infty} g(\ell)g(m)\mathbb{E}\{u(k-\ell)u(k-\tau-m)\} \\
+ \sum_{\ell=0}^{\infty} \sum_{m=0}^{\infty} h(\ell)h(m)\mathbb{E}\{e(k-\ell)e(k-\tau-m)\}
\end{aligned}
$$

$$= \sum_{\ell=1}^{\infty} \sum_{m=1}^{\infty} g(\ell)g(m)R_u(\tau + m - \ell)$$

$$+ \sum_{\ell=0}^{\infty} \sum_{m=0}^{\infty} h(\ell)h(m)R_e(\tau + m - \ell) \tag{3.49}$$

The spectrum of y can be computed from (3.40), with the covariance function $R_y(\tau)$ given by (3.49):

$$\Phi_y(\omega) = \sum_{\tau=-\infty}^{\infty} \sum_{\ell=1}^{\infty} \sum_{m=1}^{\infty} g(\ell)g(m)R_u(\tau + m - \ell)e^{-i\tau\omega}$$

$$+ \sum_{\tau=-\infty}^{\infty} \sum_{\ell=0}^{\infty} \sum_{m=0}^{\infty} h(\ell)h(m)R_e(\tau + m - \ell)e^{-i\tau\omega}$$

$$= \sum_{\ell=1}^{\infty} \sum_{m=1}^{\infty} \sum_{\tau=-\infty}^{\infty} g(\ell)e^{-i\ell\omega}g(m)e^{im\omega}R_u(\tau + m - \ell)e^{-i(\tau+m-\ell)\omega}$$

$$+ \sum_{\ell=0}^{\infty} \sum_{m=0}^{\infty} \sum_{\tau=-\infty}^{\infty} h(\ell)e^{-i\ell\omega}h(m)e^{im\omega}R_e(\tau + m - \ell)e^{-i(\tau+m-\ell)\omega}$$

$$= G(e^{i\omega})G(e^{-i\omega})\Phi_u(\omega) + H(e^{i\omega})H(e^{-i\omega})\Phi_e(\omega) \tag{3.50}$$

It follows then, that

$$\Phi_y(\omega) = \left|G(e^{i\omega})\right|^2 \Phi_u(\omega) + \left|H(e^{i\omega})\right|^2 \Phi_e(\omega) \tag{3.51}$$

Consequently, the spectrum of the output signal y is the weighted sum of the spectra of u and e, provided u and e are uncorrelated. In a similar way the cross spectrum of y and u is obtained, namely

$$\Phi_{yu}(\omega) = G(e^{i\omega})\Phi_u(\omega) \tag{3.52}$$

again under the assumption that u and e are uncorrelated.

3.7 *Examples: calculation of stochastic quantities*

In this section several examples will illustrate how to calculate stochastic properties of stochastic signals.

3.7.1 *Covariance function of a MA process*

Consider the following MA process for $v(k)$:

$$v(k) = c_0 e(k) + c_1 e(k-1) + c_2 e(k-2) \tag{3.53}$$

Note that this is a FIR representation with parameters $\{c_0, c_1, c_2\}$.
The signal $e(k)$ is discrete white noise, with zero mean and variance σ_e^2. The mean and covariance function of v, and the cross covariance of e and v, can be calculated as follows.

Mean: The mean $\bar{v}$ of $\{v(k)\}$ is calculated by taking the expectation of both sides of (3.53):

$$\mathbb{E}\{v(k)\} = c_0\mathbb{E}\{e(k)\} + c_1\mathbb{E}\{e(k-1)\} + c_2\mathbb{E}\{e(k-2)\}$$

Since $e(k)$ is Zero Mean White Noise, we have

$$\mathbb{E}\{e(k)\} = \mathbb{E}\{e(k-1)\} = \mathbb{E}\{e(k-2)\} = \bar{e} = 0$$

which results in

$$\bar{v} = 0 \tag{3.54}$$

Covariance and cross covariance: To obtain the auto- and cross-covariance functions, (3.53) is multiplied on both sides by $e(k-\tau)$ and $v(k-\tau)$, respectively. Taking expectations on both sides, the following is obtained:

$$e(k-\tau) : R_{ev}(-\tau) = c_0 R_e(\tau) + c_1 R_e(\tau-1) + c_2 R_e(\tau-2)$$
$$v(k-\tau) : R_v(\tau) = c_0 R_{ev}(\tau) + c_1 R_{ev}(\tau-1) + c_2 R_{ev}(\tau-2)$$

Taking several values for τ, the following set of equations results:

$$e(k) : R_{ev}(0) = c_0 R_e(0) + c_1 R_e(1) + c_2 R_e(2)$$
$$e(k-1) : R_{ev}(-1) = c_0 R_e(1) + c_1 R_e(0) + c_2 R_e(1)$$
$$e(k-2) : R_{ev}(-2) = c_0 R_e(2) + c_1 R_e(1) + c_2 R_e(0)$$
$$v(k) : R_v(0) = c_0 R_{ev}(0) + c_1 R_{ev}(-1) + c_2 R_{ev}(-2)$$
$$v(k-1) : R_v(1) = c_0 R_{ev}(1) + c_1 R_{ev}(0) + c_2 R_{ev}(-1)$$
$$v(k-2) : R_v(2) = c_0 R_{ev}(2) + c_1 R_{ev}(1) + c_2 R_{ev}(0)$$

The following observations can be made:

1. Since e is ZMWN, we have that $R_e(\tau) = \sigma_e^2 \Delta(\tau)$.

2. It then follows that $R_{ev}(\tau) = 0$ for $\tau < -2$.

3. From (3.53) it is clear that $v(k)$ cannot depend on future values $e(k+i)$, $i > 0$. Hence $R_{ev}(\tau) = 0$ for $\tau > 0$.

4. This implies that $R_v(\tau) = 0$ for $\tau > 2$. From the symmetry of $R_v(\tau)$ it follows that $R_v(\tau) = 0$ for $\tau < -2$ as well.

Filling this in, we obtain for the cross-covariance function

$$R_{ev}(0) = c_0 \sigma_e^2 \tag{3.55}$$
$$R_{ev}(-1) = c_1 \sigma_e^2$$
$$R_{ev}(-2) = c_2 \sigma_e^2$$
$$R_{ev}(\tau) = 0 \quad \text{for} \quad \tau \neq -2, -1, 0$$

The autocovariance function then becomes

$$R_v(0) = \left(c_0^2 + c_1^2 + c_2^2\right)\sigma_e^2 \tag{3.56}$$
$$R_v(1) = R_v(-1) = (c_0 c_1 + c_1 c_2)\sigma_e^2$$
$$R_v(2) = R_v(-2) = c_0 c_2 \sigma_e^2$$
$$R_v(\tau) = 0 \quad \text{for} \quad \tau \notin \{-2, -1, 0, 1, 2\}$$

Remark Note from (3.55) that the cross-covariance function R_{ev}, divided by σ_e^2, exactly yields the FIR parameters. This result can be generalized:
if the input signal is Gaussian white noise, then the cross-covariance function of the input and the output, divided by the variance of the input, gives the impulse response of the underlying system. ∎

3.7.2 Covariance function of an AR process

Consider the following second order AR process:

$$v(k) + a_1 v(k-1) + a_2 v(k-2) = e(k) \tag{3.57}$$

in which $e(k)$ is a Zero Mean White Noise sequence with variance σ_e^2.
The mean, autocovariance and cross covariance are calculated in the same way as in the previous paragraph.

Mean: Taking expectations on both sides of (3.57), we obtain

$$\bar{v} + a_1\bar{v} + a_2\bar{v} = \bar{e}$$

Since $\bar{e} = 0$, this results in

$$\bar{v} = 0 \tag{3.58}$$

Covariance and cross covariance: Analogously to the previous example, the auto- and cross-covariance functions are computed by multiplying (3.57) by $e(k-\tau)$ and $v(k-\tau)$, and taking expectations. This results in

$$e(k-\tau): R_{ve}(\tau) + a_1 R_{ve}(\tau-1) + a_2 R_{ve}(\tau-2) = R_e(\tau)$$
$$v(k-\tau): R_v(\tau) + a_1 R_v(\tau-1) + a_2 R_v(\tau-2) = R_{ve}(-\tau)$$

For different values of τ, the following set of equations is obtained:

$$e(k): R_{ve}(0) + a_1 R_{ve}(-1) + a_2 R_{ve}(-2) = R_e(0)$$
$$e(k-1): R_{ve}(1) + a_1 R_{ve}(0) + a_2 R_{ve}(-1) = R_e(1)$$
$$e(k-2): R_{ve}(2) + a_1 R_{ve}(1) + a_2 R_{ve}(0) = R_e(2)$$
$$v(k): R_v(0) + a_1 R_v(1) + a_2 R_v(2) = R_{ve}(0)$$
$$v(k-1): R_v(1) + a_1 R_v(0) + a_2 R_v(1) = R_{ve}(-1)$$
$$v(k-2): R_v(2) + a_1 R_v(1) + a_2 R_v(0) = R_{ve}(-2)$$

Again some observations can be made:

1. Since e is ZMWN, we have that $R_e(\tau) = \sigma_e^2 \Delta(\tau)$.

2. From (3.57) it is clear that $v(k)$ cannot depend on future values $e(k+i)$, $i > 0$. Hence $R_{ve}(\tau) = 0$ for $\tau < 0$.

Using these observations, we obtain

$$R_{ve}(0) = \sigma_e^2$$
$$R_{ve}(1) + a_1 R_{ve}(0) = 0$$
$$R_{ve}(2) + a_1 R_{ve}(1) + a_2 R_{ve}(0) = 0$$

which results in

$$R_{ve}(0) = \sigma_e^2 \tag{3.59}$$
$$R_{ve}(1) = -a_1\sigma_e^2$$
$$R_{ve}(2) = \left(a_1^2 - a_2\right)\sigma_e^2$$
$$R_{ve}(\tau) = -a_1 R_{ve}(\tau-1) - a_2 R_v e(\tau-2) \quad \text{for} \quad \tau > 2$$
$$R_{ve}(\tau) = 0 \quad \text{for} \quad \tau < 0$$

With these results the autocovariance function is computed:

$$R_v(0) + a_1 R_v(1) + a_2 R_v(2) = \sigma_e^2$$
$$a_1 R_v(0) + (1+a_2)R_v(1) = 0$$
$$a_2 R_v(0) + a_1 R_v(1) + R_v(2) = 0$$

from which we obtain

$$R_v(0) = \frac{1+a_2}{1+a_2-a_1^2+a_1^2 a_2-a_2^3-a_2^2}\sigma_e^2 \tag{3.60}$$

$$R_v(1) = R_v(-1) = \frac{-a_1}{1+a_2-a_1^2+a_1^2 a_2-a_2^3-a_2^2}\sigma_e^2$$

$$R_v(2) = R_v(-2) = \frac{a_1^2-a_2^2-a_2}{1+a_2-a_1^2+a_1^2 a_2-a_2^3-a_2^2}\sigma_e^2$$

$$R_v(\tau) = R_v(-\tau) = -a_1 R_v(\tau-1) - a_2 R_v(\tau-2) \quad \text{for} \quad \tau > 2$$

3.7.3 *Approximation of mean and covariance*

In Section 3.4.1 we pointed out that the mean and covariance must be estimated from a finite number of samples. In this example, the mean and covariance of a stochastic process are computed theoretically. They are compared with the estimates obtained from a finite data set. The lengths of these data sets (N) are 50, 200 and 500.

Consider the following first order AR-process:

$$v(k) + 0.5v(k-1) = e(k) \tag{3.61}$$

where $e(k)$ is a ZMWN sequence with variance 1.
The covariance function $R_v(\tau)$ can be calculated analytically, which gives:

$$R_v(0) = \frac{1}{1-0.5^2} = \frac{4}{3}$$
$$R_v(\tau) = -0.5 R_v(\tau-1) \qquad \tau > 0 \tag{3.62}$$
$$R_v(\tau) = R_v(-\tau) \qquad \forall \tau$$

The analytic covariance function, and the estimates with 50, 200 and 500 samples, are shown in Figure 3.11.

In Figure 3.11 the stars ('*') denote the exact covariance function. The '×'-signs show the covariance function, calculated from 500 samples. The 'o'-signs show the covariance function, calculated from 200 samples, and, finally, the '+'-signs give the covariance function, when only 50 samples are used for estimation. Note that the covariance function is symmetric.

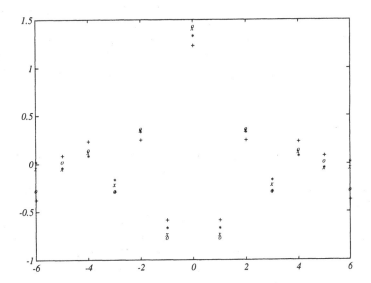

Figure 3.11 Exact (∗) and estimated covariance function: ×: $N = 500$, o: $N = 200$, +: $N = 50$

This example shows that the covariance function can be approximated fairly well from a finite number of samples. Of course, the more samples are used, the better is the approximation. One must therefore be careful when estimating the covariance function: the number of samples must not be too small.

3.8 Summary

In this chapter several mathematical model representations have been discussed. The impulse response and the transfer function appear to be useful nonparametric representations.

Disturbances have been discussed, and definitions have been given of statistical and stochastic properties of stochastic signals. A complete model can now be built, consisting of a deterministic part $G(q)$ and a stochastic part $H(q)$ (see Figure 3.8).

Next, some different parametric time-domain model structures have been mentioned. Differing mainly in the noise model, we have distinguished FIR, ARX, ARMAX, OE and BJ models.

Frequency-domain representations discussed were the frequency function, usually given in a graph, and the signal spectrum. After giving a definition of the spectrum which included both deterministic and stochastic signals, the relationship between input, output and cross spectra was established.

The calculation of stochastic properties for stochastic processes has been demonstrated by a number of examples.

Reviewing the black-box modeling procedure of Figure 3.1, we see that the next step in the procedure is the choice of an identification method: how to obtain the parameters of the model. Two types of model representations have been considered: parametric and nonparametric. Depending on the model representation that is chosen, an identification method is selected.

The next chapter will deal with nonparametric identification methods. That is, how do we obtain the parameter values of a nonparametric model of a process? The identification of parametric models is treated in Chapter 5.

3.9 References

The topics, discussed in this chapter, have been treated by a large number of authors. We will limit ourselves to mentioning only a few of them, knowing that we can never be complete. However, the references will help in finding more information on model representations.

The sampling theorem is originally presented by Shannon (1949). More about sampled-data systems, and the aspects of sampling continuous-time systems, can be found, for example, in (Åström and Wittenmark, 1984).

A more thorough discussion of system representations, such as the impulse response and the transfer function, is given by Kwakernaak and Sivan (1991). They also treat different types of stability, and they give an extensive treatment of the Fourier transform.

For stochastic theory, probability distributions, and a more thorough discussion of stationarity and ergodicity, we refer to Jenkins and Watts (1969), Beck and Arnold (1977), Caines (1988) and Hannan and Deistler (1988).

The time-domain and frequency-domain representations, as discussed here, have been presented by Ljung (1987). These parametrizations have been treated by Söderström and Stoica (1988) as well.

More about the Fourier transform and the signal spectrum can be found in (Priestley, 1989).

.10 Problems

1. Consider the following stochastic ARMA process:

$$v(k) + av(k-1) = e(k) + ce(k-1)$$

where $e(k)$ is a ZMWN sequence with variance σ_e^2.

Compute the mean value of $v(k)$, and derive expressions for the auto-covariance function of v, and for the cross-covariance function of v and e.

2. Consider the following ARMAX process:

$$y(k) + ay(k-1) = bu(k-1) + e(k) + ce(k-1)$$

where $u(k)$ and $e(k)$ are independent ZMWN sequences with variances σ_u^2 and σ_e^2, respectively.

Compute the mean value of $y(k)$, and derive expressions for the auto-covariance function of y, and for the cross-covariance function of y and e.

3. Consider the following process:

$$w(k) + fw(k-1) = bu(k-1)$$
$$v(k) + dv(k-1) = e(k) + ce(k-1)$$
$$y(k) = w(k) + v(k)$$

where $e(k)$ is a ZMWN sequence with variance σ_e^2, and $u(k) = A\sin(\omega_0 k)$. Compute the spectrum of u, the spectrum of y and the cross spectrum of y and u.

chapter four

Nonparametric identification

In the previous chapter we discussed several nonparametric model representations: impulse response, transfer function, frequency function and power
spectrum. Each model representation contains a large number of parameters.
After having chosen a model representation, the parameters must be estimated.
In this chapter the estimation of the parameters of nonparametric models is
discussed.

Identification methods are provided, with which the (large number of) parameters of the impulse response and the frequency function can be estimated.
First a time-domain method, based on correlation analysis, is presented. Then
three methods of identification of the frequency function are discussed: frequency response analysis, correlation analysis and Fourier analysis. The chapter concludes with a discussion of the demands that are put on input signals.
To extract parameters from input/output data, the input signal must be rich
enough: it must excite the system sufficiently, so as to provide the information
that is necessary for parameter identification. The notion of "richness" of a
signal is called *persistence of excitation*, to be explained in Section 4.5.

4.1 *Time domain identification by correlation analysis*

An estimate of the impulse response of a process can be obtained by correlation
analysis. The basic idea is to suppress the noise term by correlating the data
with appropriate signals.
Consider a linear time-invariant system that can be described by the following
model:

$$y(k) = \sum_{\ell=1}^{\infty} g_0(\ell)u(k-\ell) + v(k) \tag{4.1}$$

Suppose that the input u is a stationary process with zero mean, and that u is
uncorrelated with the disturbance v:

$$R_{uv}(\tau) = \mathbb{E}\{u(k)v(k-\tau)\} = 0 \qquad \forall \tau \tag{4.2}$$

Then we have

$$R_{yu}(\tau) = \mathbb{E}\{y(k)u(k-\tau)\} = \sum_{\ell=1}^{\infty} g_0(\ell)R_u(\tau-\ell) \qquad \forall \tau \tag{4.3}$$

This is an infinite set of equations for $\tau \in [1, \infty)$. However, if the system is stable, the impulse weights $g_0(\ell)$ will be zero for $\ell > M$, for some positive constant M. Then (4.3) can be written in matrix form for $\tau \in [1, M]$:

$$
\begin{bmatrix}
R_u(0) & \cdots & R_u(M-1) \\
R_u(-1) & \cdots & R_u(M-2) \\
\vdots & \ddots & \vdots \\
R_u(1-M) & \cdots & R_u(0)
\end{bmatrix}
\begin{bmatrix}
g_0(1) \\
g_0(2) \\
\vdots \\
g_0(M)
\end{bmatrix}
=
\begin{bmatrix}
R_{yu}(1) \\
R_{yu}(2) \\
\vdots \\
R_{yu}(M)
\end{bmatrix}
\tag{4.4}
$$

One clearly recognizes the covariance matrix $\mathcal{R}_u^M$ (cf. (3.19)) on the left-hand side of (4.4).

If both $R_u(\tau)$ and $R_{yu}(\tau)$ are known, a natural estimate $\hat{g}$ of the impulse response g_0 is

$$
\begin{bmatrix}
\hat{g}(1) \\
\hat{g}(2) \\
\vdots \\
\hat{g}(M)
\end{bmatrix}
= [\mathcal{R}_u^M]^{-1}
\begin{bmatrix}
R_{yu}(1) \\
R_{yu}(2) \\
\vdots \\
R_{yu}(M)
\end{bmatrix}
\tag{4.5}
$$

Of course, $R_{yu}(\tau)$ must be approximated by a finite sum. If u is known, $R_u(\tau)$ is available, and if not, $R_u(\tau)$ must be estimated as well.

A simple example will illustrate the procedure.

Example 4.1 Consider the following process:

$$
y(k) - 1.5y(k-1) + 0.7y(k-2) = u(k-1) + 0.5u(k-2) + e(k) \tag{4.6}
$$

where $e(k)$ is ZMWN with variance $\sigma_e^2 = 1$.

Suppose that $u(k)$ is selected as a ZMWN sequence with variance σ_u^2. Then the matrix $\mathcal{R}_u^M$ is a diagonal matrix, with σ_u^2 along its diagonal. Then (4.5) simplifies to

$$
\hat{g}(\ell) = \frac{R_{yu}(\ell)}{\sigma_u^2} \qquad \ell = 1, 2, \ldots, M \tag{4.7}
$$

A trial is made with $\sigma_u^2 = 1$. The results are shown in Figure 4.1. The number of samples N is 500, and $M = 40$ parameters of the impulse response $\hat{g}(\ell)$ are calculated.

The 'o'-signs in Figure 4.1 represent the true or exact impulse response. The '+'-signs form the estimated impulse response function, obtained by applying correlation analysis. The estimated impulse response is a rather good approximation of the true one. $\Box$

To solve (4.5), the matrix $\mathcal{R}_u^M$ must be invertible. Invertibility requires persistence of excitation (PE) of the signal u. This property will be discussed in Section 4.5.

Correlation analysis, as described here, is a means of obtaining the impulse response of a system. It is also possible to use it to determine the dead time of a system. This will be discussed in Section 6.2.

Another method for estimating the impulse response of a system is, e.g., that of Least Squares, which will be discussed in Chapter 5. In the following sections, we shall mention several methods of obtaining the frequency function.

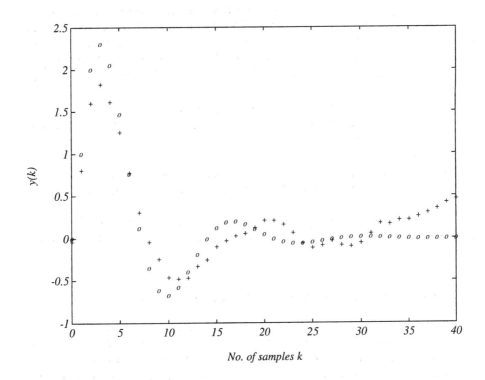

Figure 4.1 True ('o') and estimated ('+') impulse response of system (4.6)

4.2 *Frequency response analysis*

A computationally simple method to obtain an estimate of the frequency function is frequency response analysis or sine-wave testing. The idea is to apply to the system a sine wave with a specific frequency, and to measure the amplitude and the phase shift of the resulting output signal with respect to the input. The method is motivated by the fact that the frequency function $G(e^{i\omega})$ tells us what happens with a sinusoidal input: the output is a sinusoid with the same frequency; the amplitude is multiplied by $|G(e^{i\omega})|$, and the phase is shifted by $\arg[G(e^{i\omega})]$.

Consider a sinusoidal input signal (see also (3.35)–(3.37a)):

$$u(k) = A\cos(\omega_0 k) \tag{4.8}$$

If this is applied to a system with frequency function $G_0(e^{i\omega})$, the output $y(k)$ becomes

$$y(k) = A\left|G_0(e^{i\omega_0})\right|\cos(\omega_0 k + \phi) + v(k) + \text{transient} \tag{4.9}$$

where $v(k)$ is the usual noise term, and where $\phi = \arg G_0(e^{i\omega_0})$.

Neglecting the noise and the transient terms, the amplitude of the frequency function at frequency ω_0 is obtained by comparing (graphically) the amplitudes of the input $u(k)$ and the output $y(k)$. The phase shift between $u(k)$ and $y(k)$ is the phase of the frequency function at $\omega = \omega_0$.

By applying several input signals with different frequencies, and repeating the calculation steps, information about the system over a frequency range is obtained.

Frequency response analysis is widely used in practice. Amplitude and phase are usually not determined graphically, because this is very noise-sensitive. In general they are determined with curve-fitting: a sinusoid with known frequency, but unknown amplitude and phase, is fitted to the output. The amplitude and phase can be determined by nonlinear optimization, as described in Chapter 8.

The transient term can be neglected, if it has (almost) died out. Otherwise it will have a strong influence on the identification result. This, together with the fact that several input signals must be applied successively, makes the method very time-consuming.

Another disadvantage of frequency response analysis is the fact that it is not always allowed to apply purely sinusoidal input signals in industrial applications.

However, due to its simplicity the method is still widely applied, in cases where detailed information is needed about only a small number of frequencies.

4.3 *Frequency response analysis by the correlation method*

The graphical approach to frequency response analysis, as described in the previous section, is a noise-sensitive method. Due to noise it is hard to determine amplitude and phase shift graphically. In the time domain the noise is suppressed by correlation analysis. This procedure can be used in the frequency domain as well. Since the interesting component of the output $y(k)$ in (4.9) is a cosine with known frequency, correlation with a signal with the same frequency ω_0 will preserve the cosine, but will eliminate the noise term (provided there are no periodic components with frequency ω_0 present in $v(k)$).

The procedure works in the following way. Correlation of $y(k)$ with a cosine yields

$$I_C(\omega_0) = \frac{1}{N} \sum_{k=1}^{N} y(k) \cos(\omega_0 k) \tag{4.10}$$

which gives, inserting (4.9), and neglecting the transient term

$$I_C(\omega_0) = \frac{1}{N} \sum_{k=1}^{N} A \left| G(e^{i\omega_0}) \right| \cos(\omega_0 k + \phi) \cos(\omega_0 k) + \frac{1}{N} \sum_{k=1}^{N} v(k) \cos(\omega_0 k)$$

$$= A \left| G(e^{i\omega_0}) \right| \frac{1}{2} \frac{1}{N} \sum_{k=1}^{N} \left[\cos \phi + \cos(2\omega_0 k + \phi) \right] + \frac{1}{N} \sum_{k=1}^{N} v(k) \cos(\omega_0 k)$$

$$= \frac{A}{2} \left| G(e^{i\omega_0}) \right| \cos \phi + \frac{A}{2} \left| G(e^{i\omega_0}) \right| \frac{1}{N} \sum_{k=1}^{N} \cos(2\omega_0 k + \phi) +$$

$$+ \frac{1}{N} \sum_{k=1}^{N} v(k) \cos(\omega_0 k) \tag{4.11}$$

The second term on the right-hand side of (4.11) tends to zero as N tends to infinity, and so does the third term if $v(k)$ does not contain a purely periodic component of frequency ω_0.

So we obtain

$$\lim_{N \to \infty} I_C(\omega_0) = \frac{A}{2} \left| G(e^{i\omega_0}) \right| \cos \phi \qquad (4.12)$$

Correlating with a sine will give

$$I_S(\omega_0) = \frac{1}{N} \sum_{k=1}^{N} y(k) \sin(\omega_0 k) \qquad (4.13)$$

With the same assumptions as before, this becomes

$$\lim_{N \to \infty} I_S(\omega_0) = -\frac{A}{2} \left| G(e^{i\omega_0}) \right| \sin \phi \qquad (4.14)$$

Estimates of $\left| G(e^{i\omega_0}) \right|$ and ϕ are then obtained as

$$\left| \hat{G}_N(e^{i\omega_0}) \right| = \frac{\sqrt{I_C^2(\omega_0) + I_S^2(\omega_0)}}{A/2} \qquad (4.15a)$$

$$\hat{\phi}_N = \arg \hat{G}_N(e^{i\omega_0}) = -\arctan \frac{I_S(\omega_0)}{I_C(\omega_0)} \qquad (4.15b)$$

Rewriting these equations we see that

$$\hat{G}_N(e^{i\omega_0}) = \frac{I_C(\omega_0) - i I_S(\omega_0)}{A/2} \qquad (4.16)$$

The main advantage of this method is that a Bode plot can be easily obtained, and that the effort can be restricted to the frequencies of interest. The results will be better than with the frequency response analysis method, described in the previous section, since the noise is suppressed. However, the other disadvantages of frequency response analysis still apply: many industrial processes do not allow sinusoidal inputs in normal operation, steady state has to be awaited, and the experiments must be repeated a number of times (once for each frequency), which makes it a time-consuming method.

4.4 Fourier analysis: the Empirical Transfer Function Estimate

Two frequency-domain identification techniques have been described, that both have the disadvantage of long experiment periods, and requiring sinusoidal inputs. The method discussed in this section avoids these disadvantages. The method is based on (discrete) Fourier analysis of input and output signals. The resulting estimate of the frequency function is called the Empirical Transfer Function Estimate (ETFE). The ETFE will appear to be a rather rough estimate of the frequency function, and therefore techniques are presented to smooth the estimate.

4.4.1 Fourier analysis

For completeness the definitions of the Discrete Fourier Transformation (DFT) and the Inverse Discrete Fourier Transformation (IDFT) are given here.

Definition 4.1 (Discrete Fourier Transformation (DFT))
Consider the finite sequence of inputs $u(k)$, $k = 1, 2, \ldots, N$. *The Discrete Fourier Transformation maps* $u(k)$ *into* $U_N(\omega)$ *according to*

$$U_N(\omega_\ell) = \frac{1}{\sqrt{N}} \sum_{k=1}^{N} u(k) e^{-i\omega_\ell k} \qquad \omega_\ell = \frac{2\pi\ell}{N} \quad \ell = 1, \ldots, N \qquad (4.17)$$

The signal $u(k)$ can be recovered by the Inverse Discrete Fourier Transformation (IDFT) as follows.

Definition 4.2 (Inverse Discrete Fourier Transformation (IDFT))
Given $U_N(\omega)$, *as in (4.17). The Inverse Discrete Fourier Transformation maps* $U_N(\omega)$ *into* $u(k)$ *according to*

$$u(k) = \frac{1}{\sqrt{N}} \sum_{\ell=1}^{N} U_N(\omega_\ell) e^{i\omega_\ell k} \qquad \omega_\ell = \frac{2\pi\ell}{N} \qquad (4.18)$$

From (4.17) it follows that $U_N(\omega)$ is periodic with period 2π:

$$U_N(\omega + 2\pi) = U_N(\omega) \qquad (4.19)$$

Also, since $u(k)$ is real

$$U_N(-\omega) = \overline{U_N(\omega)} \qquad (4.20)$$

where the overbar denotes the complex conjugate. The function $U_N(\omega)$ is therefore completely defined by its values in the interval $[0, \pi]$. However, it is customary to consider $U_N(\omega)$ on the interval $-\pi < \omega \leq \pi$, and then (4.18) is usually written as:

$$u(k) = \frac{1}{\sqrt{N}} \sum_{\ell=-N/2+1}^{N/2} U_N(\omega_\ell) e^{i\omega_\ell k} \qquad \omega_\ell = \frac{2\pi\ell}{N} \qquad (4.21)$$

Here we have assumed that N is even. For odd N comparable summation boundaries apply.

In (4.21) we represent the signal $u(k)$ as a linear combination of $e^{i\omega_\ell k}$ for N different frequencies ω_ℓ. It is important to notice that this sum can also be written as a summation of $\cos(\omega_\ell k)$ and $\sin(\omega_\ell k)$ for the same frequencies, thus avoiding complex numbers.

Frequency response analysis by the correlation method (Section 4.3) can be seen as a special case of Fourier analysis.

Using the definition of the DFT in (4.17), we have

$$Y_N(\omega) = \frac{1}{\sqrt{N}} \sum_{k=1}^{N} y(k) e^{-i\omega k}$$

Comparing this to the definitions of I_C and I_S in (4.10) and (4.13), we see that

$$Y_N(\omega_0) = \sqrt{N} \left(I_C(\omega_0) - i I_S(\omega_0) \right)$$

Since $u(k) = A \cos(\omega_0 k)$ we have

$$U_N(\omega_0) = \frac{\sqrt{N} \, A}{2}$$

Then it follows immediately from these two equations and from (4.16), that

$$\hat{G}_N(e^{i\omega_0}) = \frac{Y_N(\omega_0)}{U_N(\omega_0)} \qquad (4.22)$$

Hence by applying Fourier analysis we arrive at the same solution as we would have obtained from correlation analysis. However, the use of Fourier analysis is not restricted to signals with one frequency only. Therefore the idea can be extended to multifrequency signals. We then arrive at the Empirical Transfer Estimate (ETFE), which will now be described.

4.4.2 Empirical Transfer Function Estimate (ETFE)

The idea of (4.22), giving an estimate of the frequency function at a certain frequency ω_0, can be easily extended to include a range of frequencies. An estimate of the transfer function is then

$$\hat{\hat{G}}_N(e^{i\omega_\ell}) = \frac{Y_N(\omega_\ell)}{U_N(\omega_\ell)} \qquad \omega_\ell = \frac{2\pi\ell}{N}, \qquad \ell = 1, \ldots, N \qquad (4.23)$$

with Y_N and U_N the Discrete Fourier Transforms of y and u, respectively. The estimate in (4.23) is called the *Empirical Transfer Function Estimate* (ETFE). This name was introduced by Ljung (1987).

The double hat denotes that the ETFE is a very rough estimate of the transfer function. This can be explained from the fact that a data series of N input and N output samples is reduced to $N/2$ complex numbers. This means that there is a relatively small data reduction.

The ETFE is illustrated by the following example.

Example 4.2 Consider the following system:

$$y(k) - 1.5y(k-1) + 0.7y(k-2) = u(k-1) + 0.5u(k-2) + e(k) \qquad (4.24)$$

where $e(k)$ is ZMWN with variance $\sigma_e^2 = 1$.
The input $u(k)$ is chosen as ZMWN with variance $\sigma_u^2 = 1$.
We generate 128 input and output samples ($N = 128$), and calculate the ETFE according to (4.23). The result is plotted in the Bode diagram of Figure 4.2.

In Figure 4.2 the solid lines correspond to the true frequency function. The dashed lines are the ETFE. It is clear that the ETFE is a very rough estimate of the frequency function. □

We mention some of the properties of the ETFE. First we establish a relation between the Discrete Fourier Transforms of the input and output of a purely deterministic system.

Theorem 4.1 *Let $y(k)$ and $u(k)$ be related by the BIBO stable system $G(q)$ as*

$$y(k) = G(q)u(k)$$

Let $u(k)$ be unknown, but bounded by a constant C_u:

$$|u(k)| \leq C_u \qquad \forall k \in \mathbb{Z}$$

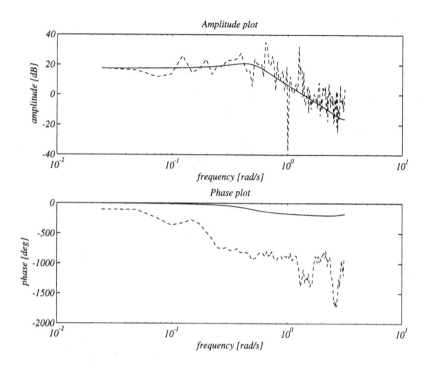

Figure 4.2 Amplitude and phase of the true system (solid) and the ETFE (dashed), based on 128 measurements

Then the Discrete Fourier Transforms $Y_N(\omega)$ of y and $U_N(\omega)$ of u are related by

$$Y_N(\omega) = G(e^{i\omega})U_N(\omega) + R_N(\omega) \tag{4.25}$$

where

$$|R_N(\omega)| \leq \frac{2.C_u.C_G}{\sqrt{N}}$$

with

$$C_G = \sum_{\ell=1}^{\infty} \ell |g(\ell)| \tag{4.26}$$

The values $g(\ell)$ are the impulse response of the system with transfer function $G(q)$.

Proof We have

$$|R_N(\omega)| = \left| Y_N(\omega) - G(e^{i\omega})U_N(\omega) \right| \tag{4.27}$$

By definition of the Discrete Fourier Transformation (4.17) we have

$$Y_N(\omega) = \frac{1}{\sqrt{N}} \sum_{k=1}^{N} y(k)e^{-i\omega k} = \frac{1}{\sqrt{N}} \sum_{k=1}^{N} \sum_{\ell=1}^{\infty} g(\ell)u(k-\ell)e^{-i\omega k}$$

$$= \frac{1}{\sqrt{N}} \sum_{\ell=1}^{\infty} g(\ell)e^{-i\omega \ell} . \sum_{m=1-\ell}^{N-\ell} u(m)e^{-i\omega m} \tag{4.28}$$

The last equality follows from substitution of $m = k - \ell$.
It follows from (4.27) and (4.28) that

$$|R_N(\omega)| \leq \left| \sum_{\ell=1}^{\infty} g(\ell)e^{-i\omega\ell} \right| \cdot \left| \frac{1}{\sqrt{N}} \sum_{m=1-\ell}^{N-\ell} u(m)e^{-i\omega m} - U_N(\omega) \right| \qquad (4.29)$$

For the second part of the right-hand side of (4.29) we have

$$\left| \frac{1}{\sqrt{N}} \sum_{m=1-\ell}^{N-\ell} u(m)e^{-i\omega m} - U_N(\omega) \right|$$

$$= \left| \frac{1}{\sqrt{N}} \sum_{m=1-\ell}^{N-\ell} u(m)e^{-i\omega m} - \frac{1}{\sqrt{N}} \sum_{k=1}^{N} u(k)e^{-i\omega k} \right|$$

$$\leq \left| \frac{1}{\sqrt{N}} \sum_{m=1-\ell}^{0} u(m)e^{-i\omega m} \right| + \left| \frac{1}{\sqrt{N}} \sum_{m=N-\ell+1}^{N} u(m)e^{-i\omega m} \right|$$

$$\leq 2 \cdot \frac{\ell.C_u}{\sqrt{N}} \qquad\qquad \ell = 1, \ldots, \infty \qquad (4.30)$$

It then follows that

$$|R_N(\omega)| \leq \left| \sum_{\ell=1}^{\infty} \ell.g(\ell)e^{-i\omega\ell} \right| \cdot \frac{2C_u}{\sqrt{N}} \qquad (4.31)$$

With C_G defined as in (4.26), this becomes

$$|R_N(\omega)| \leq \frac{2.C_u.C_G}{\sqrt{N}} \qquad (4.32)$$

∎

Note that for periodic signals $u(k)$ with period N, $R_N(\omega)$ in (4.27) is zero for $\omega = 2\pi\ell/N$, $\ell = 1, \ldots, N$. This follows from the fact that the left-hand side of (4.30) is zero for these signals.

By using Theorem 4.1, and the definition of the ETFE (4.23), the following can be established. Suppose that a system with true transfer function $G_0(q)$ satisfies (4.1). Then the ETFE, based on N measurements of y and u, satisfies

$$\hat{G}_N(e^{i\omega}) = G_0(e^{i\omega}) + \frac{R_N(\omega)}{U_N(\omega)} + \frac{V_N(\omega)}{U_N(\omega)} \qquad (4.33)$$

where $V_N(\omega)$ is the DFT of the noise $v(k)$, and $R_N(\omega)$ is defined as in Theorem 4.1.

The noise $v(k)$ is supposed to have zero mean, and hence $\mathbb{E}\{V_N(\omega)\} = 0, \forall\omega$. The *bias* in the estimate is thus $R_N(\omega)/U_N(\omega)$.

Bias and covariance of the ETFE are summarized in the following theorem.

Theorem 4.2 *Consider a BIBO stable system*

$$y(k) = G_0(q)u(k) + v(k) \qquad (4.34)$$

The disturbance $v(k)$ has zero mean, and spectrum $\Phi_v(\omega)$. Let $\{u(k)\}$ be independent of $\{v(k)\}$, and assume that $|u(k)| \leq C_u$ for all k, for some positive constant C_u. The ETFE has the following properties:

1. *The ETFE has a bias, which follows from*

$$\mathbb{E}\left\{\hat{G}_N(e^{i\omega})\right\} = G_0(e^{i\omega}) + \frac{R_N(\omega)}{U_N(\omega)} \tag{4.35}$$

with

$$|R_N(\omega)| \leq \frac{C_1}{\sqrt{N}} \tag{4.36}$$

where

$$C_1 = 2 \cdot C_u \cdot C_G \tag{4.37}$$

and C_G is defined by (4.26).

2. *The covariance of the ETFE is given by*

$$\mathbb{E}\left\{\left[\hat{G}_N(e^{i\omega}) - G_0(e^{i\omega})\right]\left[\hat{G}_N(e^{i\xi}) - G_0(e^{i\xi})\right]\right\}$$

$$= \begin{cases} \dfrac{\Phi_v(\omega) + \rho_N(\omega)}{|U_N(\omega)|^2} & \xi = \omega \\[3mm] \dfrac{\rho_N(\omega)}{U_N(\omega)U_N(-\xi)} & |\xi - \omega| = \frac{2\pi k}{N}, \quad k = 1, \ldots, N-1 \end{cases} \tag{4.38}$$

with

$$|\rho_N(\omega)| \leq \frac{C_2}{N} \tag{4.39}$$

where C_2 is defined as

$$C_2 = C_1^2 + \sum_{\tau=-\infty}^{\infty} |\tau R_v(\tau)| \tag{4.40}$$

The proof can be found in (Ljung, 1987).

From Theorem 4.2 we can draw some conclusions with respect to the quality of the estimate, obtained with Fourier analysis.

Since for periodic signals $u(k)$, $R_N(\omega) = 0$ in (4.27) and (4.35), we distinguish two cases: periodic inputs and nonperiodic (stochastic) inputs.

Periodic input: The Fourier Transform of the input only exists at a fixed number of frequencies. We have that

  1. The ETFE is an unbiased estimate of the transfer function.
  2. The variance of the ETFE decays as $1/N$.
  3. The covariance of the ETFE decays as $1/N$.

Nonperiodic input: The Fourier transform is defined at N frequencies. We have that

  1. The ETFE is an asymptotically unbiased estimate of the transfer function. Here asymptotically means that $N \to \infty$.
  2. The variance of the ETFE does not decrease as N increases. It is determined by the noise-to-signal ratio $\Phi_v(\omega)/|U_N(\omega)|^2$.

3. The covariance of the ETFE decays as $1/N$. This means that asymptotically ($N \to \infty$) the estimates at two different frequencies $\omega \neq \xi$ are uncorrelated.

From this discussion it follows that the ETFE is of increasingly good quality, if a periodic input is used and the number of samples is increased. However, if the input is not periodic, the estimate does not improve as the number of samples increases. This is because we not only increase the data length, but we also increase the number of estimated parameters. Hence the amount of information per parameter does not increase, and therefore the variance does not decrease. This makes the ETFE a rather crude estimate of the transfer function (see also Example 4.2).

To improve the estimate, use can be made of the fact that for any system there should be some correlation between two subsequent frequency points of the ETFE. This is what is done in the next subsection.

4.4.3 Smoothing of the ETFE

The ETFE can be improved by making use of the uncorrelatedness of the observations at different frequencies. A possible way to do this is by taking expectation over a number of observations. This is illustrated by the following example.

Example 4.3 Let $\{e(k)\}$ be a sequence of uncorrelated observations of a constant quantity:

$$\bar{\mathbb{E}}\{e(k)\} = c \tag{4.41}$$

$$\bar{\mathbb{E}}\{[e(k) - c][e(\ell) - c]\} = 0, \qquad k \neq \ell \tag{4.42}$$

The variance of $e(k)$ is σ_e^2:

$$\bar{\mathbb{E}}\{[e(k) - c]^2\} = \sigma_e^2$$

Let the mean of these quantities be denoted by $\hat{c}$:

$$\hat{c} = \frac{1}{N} \sum_{k=1}^{N} e(k) \tag{4.43}$$

Show that

$$\bar{\mathbb{E}}\{\hat{c}\} = c \tag{4.44}$$

$$\bar{\mathbb{E}}\{[\hat{c} - c]^2\} = \frac{\sigma_e^2}{N} \tag{4.45}$$

Solution: First we show (4.44). We have

$$\bar{\mathbb{E}}\{\hat{c}\} = \bar{\mathbb{E}}\left\{\frac{1}{N} \sum_{k=1}^{N} e(k)\right\} = \frac{1}{N} \sum_{k=1}^{N} \bar{\mathbb{E}}\{e(k)\} = \frac{1}{N} \sum_{k=1}^{N} c = c \tag{4.46}$$

For the covariance we have

$$\mathbb{E}\{[\hat{c} - c]^2\} = \mathbb{E}\left\{\left[\frac{1}{N}\sum_{k=1}^{N} e(k) - c\right]^2\right\} = \mathbb{E}\left\{\left[\frac{1}{N}\sum_{k=1}^{N}\{e(k) - c\}\right]^2\right\}$$

$$= \frac{1}{N^2}\mathbb{E}\left\{\sum_{k=1}^{N}[e(k) - c]^2\right\} = \frac{1}{N^2}\sum_{k=1}^{N}\mathbb{E}\left\{[e(k) - c]^2\right\} \qquad (4.47)$$

where the last equality is only possible because the observations are uncorrelated according to (4.42). Then (4.45) follows easily. ∎

From this example we see that the ETFE at frequency ω_0 can be improved by averaging over a frequency interval. This is allowed if the following conditions are satisfied:

1. The true transfer function $G_0(e^{i\omega})$ is a smooth function of ω

2. The frequency distance $2\pi/N$ is small compared to how quickly $G_0(e^{i\omega})$ changes

Under these conditions the ETFE $\hat{G}_N(e^{i\omega})$ at the frequencies ω in the interval

$$\frac{2\pi k_1}{N} = \omega_0 - \Delta\omega \leq \omega \leq \omega_0 + \Delta\omega = \frac{2\pi k_2}{N} \qquad (4.48)$$

can be used to improve the estimate at ω_0 as

$$\hat{G}_N(e^{i\omega_0}) = \frac{1}{k_2 - k_1 + 1}\sum_{\ell=k_1}^{k_2}\hat{G}_N(e^{i\omega_\ell}) \qquad \omega_\ell = \frac{2\pi\ell}{N} \qquad (4.49)$$

The resulting estimate is the average of the ETFE over a window with width $k_2 - k_1 + 1$.

Note that if the width of the window is large, the transfer function is not constant over the frequency interval. This will introduce a bias in the resulting estimate. In summary we can say that averaging over a window introduces (or enlarges) a bias, and reduces the variance. The larger the window is, the larger are its effects. Consequently, the choice of the width of the window is a trade-off between bias and variance. This is illustrated by the next example.

Example 4.4 Consider the same system as in Example 4.2:

$$y(k) - 1.5y(k-1) + 0.7y(k-2) = u(k-1) + 0.5u(k-2) + e(k) \qquad (4.50)$$

where $e(k)$ is ZMWN with variance $\sigma_e^2 = 1$.
The input $u(k)$ is chosen as ZMWN with variance $\sigma_u^2 = 1$.

The same data is used as in Example 4.2 ($N = 128$). In Figure 4.3 the amplitude and phase of the real system and the smoothed estimates are shown. The solid line is the real system. The ETFE of Figure 4.2 is smoothed with two different Hamming windows, one of width 5 (dotted line), and the other of width 10 (dash-dotted line).

It can be seen that the windows both give a smoother estimate than the raw ETFE. Especially in the low-frequency region, where the frequency function is

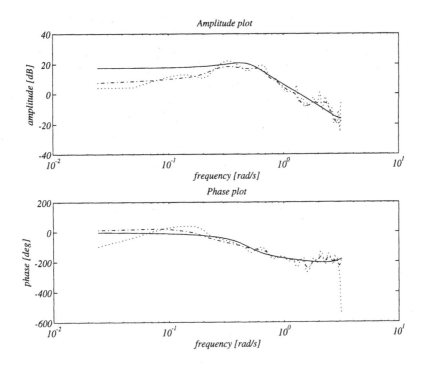

Figure 4.3 Amplitude and phase of the true system (solid) and the smoothed ETFE with a window width of 5 (dotted) and 10 (dash-dotted)

almost constant, the noise is completely filtered out. For higher frequencies the variance has decreased as well, but the bias has increased. As expected, the wider window (dash-dotted) gives a larger bias than the narrower window (dotted).   ☐

The above example clearly demonstrates that the ETFE can be smoothed by filtering with a window. The width of the window must be chosen with care, and is a trade-off between smaller variance and larger bias.

4.5 Persistence of Excitation

As we have already indicated, the input $u(k)$ must be "rich" enough for identification. The notion of *persistence of excitation* formalizes this idea.

If we carry out an experiment to obtain the parameters of an unknown process, we apply an input signal $u(k)$ to the process and measure the output $y(k)$. The data set $\{u(k), y(k)\}_N$ is subsequently interpreted by a parameter estimation algorithm to estimate the parameters θ of a model. In general, we assume that the input $u(k)$ is "rich" enough to excite all required parameters of the process such that their values are reflected in the output $y(k)$.

If we select $u(k) = \sin(\omega_0 k)$, the output becomes $y(k) = A\sin(\omega_0 k + \phi)$ with $A(\theta)$ the amplitude and $\phi(\theta)$ the phase at $\omega = \omega_0$ of the frequency function of the process. Each frequency ω_i yields a value $A(\theta, \omega_i)$ and a value of the phase $\phi(\theta, \omega_i)$. Consequently, with the selection of $u(k) = \sin(\omega_0 k)$, two equations are generated for solving the unknown parameters θ. If the model of the process has only two parameters, both parameters can be solved for. If more than two parameters have to be estimated, this selection of $u(k)$ is not appropriate. By selecting an input $u(k)$ as the sum of n different frequencies, $2n$ equations are produced, so $2n$ parameters can be estimated.

This example illustrates that if the input $u(k)$ becomes "richer", more parameters can be estimated. If the input is not "rich" enough, not all parameters can be found. The "richness" of $u(k)$ depends on its frequency contents, as reflected in its spectrum $\Phi_u(\omega)$. The "richness" of $u(k)$ can be formulated mathematically as the degree of *persistence of excitation*, abbreviated as PE, of $u(k)$.

Definition 4.3 (Persistence of Excitation)
A quasi-stationary signal $u(k)$ with spectrum $\Phi_u(\omega)$ is persistently exciting of order n if, for all moving average filters of the form

$$M_n(q) = m_1 q^{-1} + \ldots + m_n q^{-n} \tag{4.51}$$

the relation

$$|M_n(e^{i\omega})|^2 \Phi_u(\omega) \equiv 0 \tag{4.52}$$

implies that $M_n(e^{i\omega}) \equiv 0$.

This formal definition can be interpreted as follows.

- The input $u(k)$ is persistently exciting of order n if $\Phi_u(\omega)$ is different from zero at at least n points in the interval $-\pi < \omega \le \pi$.

or

- A signal $u(k)$ with degree n of persistence of excitation cannot be suppressed completely by an $(n-1)^{\text{th}}$ order moving average filter.

A consequence of the order n of persistence of excitation of $u(k)$ is that with this $u(k)$ a maximum of n parameters can be estimated. Hence, $u(k) = \sin(\omega_0 k)$ with a degree two of persistence of excitation, can estimate a maximum of two parameters. The sum of three sinusoids with different frequencies enables the estimation of a maximum of six parameters. A square wave , which has many different values of its spectrum not equal zero, allows the estimation of many different parameters.

Another consequence of the definition of PE is that the covariance matrix $\mathcal{R}_u^n$ of $u(k)$ is nonsingular if $u(k)$ has at least a degree n of persistence of excitation.

4.6 *Summary*

In this chapter several nonparametric identification methods have been discussed, both in the time domain and in the frequency domain. In Table 4.1 their principles and characteristics are summarized.

As illustrated in Example 4.4 it is possible to improve the ETFE by averaging the ETFE over a window. The width of the window must be chosen as a trade-off between bias and variance of the resulting estimate.

Finally the concept of Persistence of Excitation has been discussed. A mathematical definition has been provided, but it is probably easier to recall that to estimate a number of n parameters, the input must be persistently exciting (PE) of order n.

Table 4.1
Nonparametric identification methods

Method	Domain	Principle	Characteristics
Correlation	time	correlation of output with input, and solve for the impulse response	• each input signal possible • approximation of the covariance
Frequency response	frequency	apply a sine wave, and determine amplitude and phase graphically or by optimization	• sinusoid inputs only • long experiment period • transient effects
Frequency response + correlation	frequency	apply a sine wave, correlate with sine and cosine	• sinusoid inputs only • long experiment period • transient effects
Fourier analysis (ETFE)	frequency	ETFE = DFT of output / DFT of input	• each input signal possible • fast • noise-sensitive • can be smoothed

4.7 *References*

There is a vast amount of literature dedicated to the topics, treated in this chapter. Of these references, we can but mention a few.

Correlation analysis in the time domain and its implications are treated by Graupe (1972). The frequency-domain approach is described by Jenkins and Watts (1969), and by Priestley (1989).

A more thorough treatment of both time- and frequency-domain methods, and of the concept of persistence of excitation (PE), is given by Ljung (1987) and by Söderström and Stoica (1989).

Finally we mention the Reference Guide of the Identification Toolbox, available with Matlab (Ljung, 1991). The Toolbox provides several nonparametric identification methods, such as spectral analysis and ETFE calculation. In the Reference Guide these methods are documented.

4.8 Problems

1. Consider the input signal

$$u(k) = A\sin(\omega_0 k)$$

Construct the covariance matrices $\mathcal{R}_u^1$, $\mathcal{R}_u^2$ and $\mathcal{R}_u^3$, and determine whether they are singular.

How many parameters can be estimated with this u?

chapter five

Parametric identification

Parametric identification is concerned with the identification of parametric models. Parametric models can be used for process analysis, for simulation and for control design.

The parametric identification methods described in this chapter are based on solving a minimization problem. A model structure is chosen, and the parameters in the structure are obtained by minimizing a cost function. The idea is explained in Section 5.1. It will be seen that the methods use a prediction model, and in Section 5.2 the prediction models are derived for the model structures, discussed in Section 3.5. The solution method, that of Least Squares, is discussed in Section 5.3. Of course, it is very important to know the quality of a model, and in Section 5.4 this problem is addressed in the case of the linear Least Squares estimate (ARX- and FIR-models). In Sections 5.5 and 5.6 the discussion is extended to include other more general models (ARMAX, OE). The chapter concludes with a summary.

5.1 Prediction error methods

Obtaining a model is never a goal in itself. There is always a purpose in obtaining a model. In many cases the purpose is process analysis and/or control design. In process analysis we want to predict the process behavior in certain circumstances. Usually the analysis is done by simulation. In control design we want to develop a controller, which makes the process behave in a desired way.

In both cases the model must be able to predict the behavior of the process. Therefore a very logical criterion for model quality is its prediction capacity. How does one measure the capability of a model to predict the process behavior?

A possible choice is to calculate a future output, and compare this with the actual output. If the error between measured and predicted output is small enough in some sense, the model can be considered good. Methods that are based on this principle are called Prediction Error Methods (PEMs), since the prediction error (error between measured and predicted output) is used as a measure of quality.

To formalize the concept, let us introduce some notation. Suppose $y(k)$ denotes the measured process output at instant $t = k$. Suppose that we have complete knowledge of the past at $t = k$ (through measurements), and that we want to predict the process output at $t = k + 1$. The predicted process output is then denoted by $\hat{y}(k+1|k)$, which is called the *one-step ahead prediction* of y. At time instant $k+1$ the value of $y(k+1)$ is measured. Comparing the predicted process output $\hat{y}(k+1|k)$ with the measured process output $y(k+1)$, we obtain the prediction error $\epsilon(k+1|k)$:

$$\epsilon(k+1|k) = y(k+1) - \hat{y}(k+1|k) \tag{5.1}$$

According to the philosophy described above, a model is good if it has a small prediction error. But for the same model, the prediction error can be small at one instant, and large at another. Therefore a much better measure would be to look at a sequence of prediction errors. To prevent positive and negative prediction errors from cancelling out, the square can be taken. Then an appropriate measure for model quality is the following loss function J_N:

$$J_N = \frac{1}{N} \sum_{k=1}^{N} \epsilon^2(k|k-1) \tag{5.2}$$

A model has good prediction capabilities, if the loss function (5.2) is small enough. The "best" model is thus the model that minimizes (5.2). The identification problem is formulated as a minimization problem.

As was stated before, a parametric model consists of a model structure (ARX, ARMAX, etc.), a set of integers (dead time, polynomial orders), and a number of parameters. Suppose that these parameters are collected in the *parameter vector* θ (this will be further explained in the next section). Then the optimal model within the selected model structure is the model with the parameter vector $\hat{\theta}_N$, that minimizes (5.2):

$$\hat{\theta}_N = \arg \min_{\theta \in D_M} J_N(\theta) \tag{5.3}$$

with

$$J_N(\theta) = \frac{1}{N} \sum_{k=1}^{N} \epsilon^2(k|k-1, \theta) \tag{5.4}$$

where the parameter-dependency in J_N and ϵ has been made more explicit. The set D_M is the set of possible parameter vectors for a chosen model parametrization. This set is not made more explicit here, but one could think for example of the set of parameters which constitute a second order ARX model.

Prediction Error Methods are based on (5.3) and (5.4). In the following section they are worked out for the different model structures introduced in Section 3.5.

5.2 *Prediction models*

From (5.1) it is clear that, to compute the prediction error, a prediction of y is needed. In this section the problem of one-step ahead prediction is addressed. First this is done for a general process and noise model. Then attention is focused on the model structures that were presented in Section 3.5.

5.2.1 *One-step ahead prediction*

Prediction Error Methods are based on the one-step ahead prediction of the process output. How can the one-step ahead prediction $\hat{y}(k+1|k)$ be calculated?

To answer this question, we go back to our standard model, Figure 3.8. For convenience it is shown again in Figure 5.1.

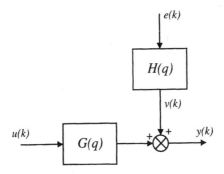

Figure 5.1 Standard model

The process output is described by

$$y(k) = G(q)u(k) + H(q)e(k) = G(q)u(k) + v(k) \tag{5.5}$$

In (5.5) we know $\{y(\ell), u(\ell)\}, \ell = k - 1, k - 2, \ldots$ from measurements. For the moment suppose that $G(q)$ and $H(q)$ are known as well. Then also $v(\ell)$ is known for $\ell = k - 1, k - 2, \ldots$, according to

$$v(\ell) = y(\ell) - G(q)u(\ell) \qquad \ell = k - 1, k - 2, \ldots \tag{5.6}$$

The one-step ahead prediction of y is given by:

$$\hat{y}(k|k - 1) = G(q)u(k) + \hat{v}(k|k - 1) \tag{5.7}$$

Note that, since $G(q)$ is known and strictly causal, $G(q)u(k)$ only contains values up to $k - 1$, and hence it is known at time instant $k - 1$.

To solve the prediction problem, first the problem of obtaining $\hat{v}(k|k - 1)$ must be treated. To make the problem soluble, the noise filter $H(q)$ must satisfy three conditions. This will become clear later in the discussion. The three conditions are:

1. $H(q)$ is stable

2. $H(q)$ is inversely stable, i.e., $H^{-1}(q) = 1/H(q)$ is stable

3. $H(q)$ is monic.

The third condition implies that the noise process can be written in its impulse response representation as

$$v(k) = e(k) + \sum_{\ell=1}^{\infty} h(\ell)e(k - \ell) \tag{5.8}$$

The question now is, how to predict $v(k)$ from past measurements of v? In other words, if $v(\ell)$ is known for $\ell \leq k - 1$, what is the best prediction of $v(k)$?

The best prediction is, of course, the expectation of (5.8). We then obtain

$$\hat{v}(k|k - 1) = \mathbb{E}\{v(k)\} = \mathbb{E}\{e(k)\} + \mathbb{E}\left\{\sum_{\ell=1}^{\infty} h(\ell)e(k - \ell)\right\}$$

Since e is a ZMWN sequence, the first part is zero. The second part is deterministic, since we assumed that $v(\ell)$, and hence $e(\ell)$, is known for $\ell \leq k - 1$. Hence the prediction becomes

$$\hat{v}(k|k-1) = \sum_{\ell=1}^{\infty} h(\ell)e(k-\ell) = \left[H(q) - 1\right] e(k)$$

If the noise process is inversely stable (condition 2), we can write

$$e(k) = \frac{1}{H(q)} v(k)$$

which gives

$$\hat{v}(k|k-1) = \frac{H(q) - 1}{H(q)} v(k) \tag{5.9}$$

It is easily verified that to compute the above prediction of $v(k)$ only values of $v(\ell)$ for $\ell \leq k - 1$ are needed.

Now that we have (5.9), we can reconsider (5.7) for the one-step ahead prediction of y:

$$\hat{y}(k|k-1) = G(q)u(k) + \left[1 - H^{-1}(q)\right] v(k) \tag{5.10}$$

where measurements of $v(\ell)$ for $\ell \leq k - 1$ are obtained from (5.6). Then (5.10) becomes

$$\hat{y}(k|k-1) = G(q)u(k) + \left[1 - H^{-1}(q)\right] \left[y(k) - G(q)u(k)\right]$$

which can be rewritten as

$$\hat{y}(k|k-1) = H^{-1}(q)G(q)u(k) + \left[1 - H^{-1}(q)\right] y(k) \tag{5.11}$$

or as

$$H(q)\hat{y}(k|k-1) = G(q)u(k) + \left[H(q) - 1\right] y(k)$$

With (5.11) the prediction error (5.1) can be written as

$$\epsilon(k|k-1) = y(k) - \hat{y}(k|k-1) = -H^{-1}(q)G(q)u(k) + H^{-1}(q)y(k) \tag{5.12}$$

The above derivation is valid for any model structure that can be written as (5.5). Therefore (5.11) is a very general expression for the one-step ahead prediction of y.

In the derivation of (5.12) we have assumed that $G(q)$ and $H(q)$ were known. This is not true, of course, since finding these transfer functions is the goal of identification. If they are known, there is no need for identification.

To take this into account, both $G(q)$ and $H(q)$ are *parametrized* with a parameter vector θ, as will be shown in the next sections. Then (5.12) becomes

$$\epsilon(k|k-1,\theta) = H^{-1}(q,\theta)[y(k) - G(q,\theta)u(k)] \tag{5.13}$$

By varying θ over the domain $D_{\mathcal{M}}$, we can compute ϵ, and calculate the associated value of J_N in (5.4).

In Chapter 3 several model structures were discussed. Each model structure described in Section 3.5 has its own dedicated prediction model, with certain properties. This will be discussed in the following subsections.

5.2.2 Prediction model for the FIR structure

The FIR model structure is given by

$$y(k) = B(q)u(k) + e(k) \tag{5.14}$$

The polynomial $B(q)$ has parameters $\{b_1, \ldots, b_{n_b}\}$. For easy reference these parameters are collected in a parameter vector θ:

$$\theta = \begin{pmatrix} b_1 & b_2 & \cdots & b_{n_b} \end{pmatrix}^T \tag{5.15}$$

Comparing (5.14) to (5.5) we see that

$$G(q, \theta) = B(q) \qquad H(q, \theta) = 1 \tag{5.16}$$

The parameter vector θ is mentioned explicitly in the transfer functions, to show the dependency of the model on the parameters. The one-step ahead prediction of a FIR model then becomes, using (5.11):

$$\hat{y}(k|\theta) = B(q)u(k) \tag{5.17}$$

For obvious reasons (5.17) is called a FIR *prediction model*.
The measurements of $u(\ell)$ for $\ell \le k - 1$ can be collected in a so-called *regression vector* ϕ:

$$\phi(k) = \begin{pmatrix} u(k-1) & u(k-2) & \cdots & u(k-n_b) \end{pmatrix}^T \tag{5.18}$$

With (5.15) and (5.18) the prediction model (5.15) can be written in *regression form*:

$$\hat{y}(k|\theta) = \theta^T \phi(k) = \phi^T(k)\theta \tag{5.19}$$

Since $\hat{y}(k|\theta)$ is linearly dependent on θ, this is called a *linear regression model*.

Remark A linear regression model does not necessarily have linear dynamics (input-output behavior). Certain types of nonlinear dynamic models can be described by linear regression models as well. ∎

5.2.3 Prediction model for the ARX structure

The ARX model structure is given by

$$A(q)y(k) = B(q)u(k) + e(k) \tag{5.20}$$

Collect the parameters a_i and b_i in the parameter vector θ:

$$\theta = \begin{pmatrix} a_1 & a_2 & \cdots & a_{n_a} & b_1 & \cdots & b_{n_b} \end{pmatrix}^T \tag{5.21}$$

From (5.20) and (5.5) it follows that

$$G(q, \theta) = \frac{B(q)}{A(q)} \qquad H(q, \theta) = \frac{1}{A(q)} \tag{5.22}$$

Using (5.11), the ARX prediction model becomes

$$\hat{y}(k|\theta) = B(q)u(k) + \left[1 - A(q)\right] y(k) \tag{5.23}$$

If the regression vector $\phi(k)$ is defined as

$$\phi(k) = \begin{pmatrix} -y(k-1) & -y(k-2) & \cdots & -y(k-n_a) & u(k-1) & \cdots \\ \cdots & u(k-n_b) \end{pmatrix}^T$$

$$(5.24)$$

the ARX prediction model can be written as a linear regression model:

$$\hat{y}(k|\theta) = \theta^T \phi(k) = \phi^T(k)\theta \tag{5.25}$$

5.2.4   Prediction model for the ARMAX structure

The ARMAX model structure is given by

$$A(q)y(k) = B(q)u(k) + C(q)e(k) \tag{5.26}$$

Again the parameters of the A-, B- and C- polynomials are collected in the parameter vector θ:

$$\theta = \begin{pmatrix} a_1 & \cdots & a_{n_a} & b_1 & \cdots & b_{n_b} & c_1 & \cdots & c_{n_c} \end{pmatrix}^T \tag{5.27}$$

From (5.5) it follows that

$$G(q,\theta) = \frac{B(q)}{A(q)} \qquad H(q,\theta) = \frac{C(q)}{A(q)} \tag{5.28}$$

The ARMAX prediction model follows from (5.11):

$$\hat{y}(k|\theta) = \frac{B(q)}{C(q)}u(k) + \left[1 - \frac{A(q)}{C(q)}\right]y(k) \tag{5.29}$$

The ARMAX prediction model can also be written in a regression form. Multiply (5.29) with $C(q)$ and add $\left[1 - C(q)\right]\hat{y}(k|\theta)$ to both sides, which yields

$$\hat{y}(k|\theta) = B(q)u(k) + \left[C(q) - A(q)\right]y(k) + \left[1 - C(q)\right]\hat{y}(k|\theta)$$

which can be rewritten as

$$\hat{y}(k|\theta) = B(q)u(k) + \left[1 - A(q)\right]y(k) + \left[C(q) - 1\right]\left[y(k) - \hat{y}(k|\theta)\right] \tag{5.30}$$

Using the definition of the prediction error $\epsilon(k,\theta)$ in (5.1),

$$\epsilon(k|\theta) = y(k) - \hat{y}(k|\theta)$$

(5.30) can be rewritten as

$$\hat{y}(k|\theta) = B(q)u(k) + \left[1 - A(q)\right]y(k) + \left[C(q) - 1\right]\epsilon(k|\theta) \tag{5.31}$$

Select the following regression vector $\phi(k)$ for the ARMAX prediction model:

$$\phi(k|\theta) = \begin{pmatrix} -y(k-1) & \cdots & -y(k-n_a) & u(k-1) & \cdots & u(k-n_b) \\ \epsilon(k-1|\theta) & \cdots & \epsilon(k-n_c|\theta) \end{pmatrix}^T$$

$$(5.32)$$

Note that the regression vector has become a function of the parameter vector θ, since it contains the θ-dependent prediction errors.

The ARMAX prediction model can now be written in the following regression form, using (5.27), (5.31) and (5.32):

$$\hat{y}(k|\theta) = \theta^T \phi(k|\theta) = \phi^T(k|\theta)\theta \tag{5.33}$$

The similarity between (5.33) and (5.25) is clear. Because $\hat{y}(k|\theta)$ is not a linear function of θ in (5.33), the ARMAX prediction model is called a *pseudo-linear regression model*.

5.2.5 Prediction model for the OE structure

The Output Error model is given by

$$F(q)w(k) = B(q)u(k) \tag{5.34a}$$
$$y(k) = w(k) + e(k) \tag{5.34b}$$

As in the previous subsections, the process and noise models are clear:

$$G(q, \theta) = \frac{B(q)}{F(q)} \qquad H(q, \theta) = 1 \tag{5.35}$$

The parameter vector θ contains the parameters of the B- and F-polynomials:

$$\theta = \begin{pmatrix} b_1 & \cdots & b_{n_b} & f_1 & \cdots & f_{n_f} \end{pmatrix}^T \tag{5.36}$$

The OE prediction model follows easily from (5.11):

$$\hat{y}(k|\theta) = \frac{B(q)}{F(q)}u(k) = w(k|\theta) \tag{5.37}$$

Note that the signal w is not observed, and that it can only be reconstructed using (5.34a). Therefore it is θ-dependent.

The prediction model is obtained by defining the following regression vector:

$$\phi(k|\theta) = \begin{pmatrix} u(k-1) & \cdots & u(k-n_b) \\ -w(k-1|\theta) & \cdots & -w(k-n_f|\theta) \end{pmatrix}^T \tag{5.38}$$

The regression model is then again

$$\hat{y}(k|\theta) = \theta^T \phi(k|\theta) = \phi^T(k|\theta)\theta \tag{5.39}$$

This prediction model exhibits again a nonlinear function $\hat{y}(k|\theta)$ of θ, so it is a pseudo-linear regression model.

The prediction models, derived in this section, will be used for identification with a Prediction Error Method. This will be further elaborated on in the next section.

5.3 Least Squares method

As was pointed out in Section 5.1, the parametric identification methods that are considered belong to the family of Prediction Error Methods (PEM). It was explained that these methods are based on the minimization of a cost function, which is the sum of the squared prediction errors. To solve this minimization problem, use is made of mathematical techniques. One such technique is the Least Squares (LS) method. In this section the LS method is explained, and it is shown how the parameters of a model can be found by such a method. This is also called *parameter estimation*. First the simplest case is considered, in which the model is a linear regression model (FIR, ARX). The more complicated model structures, such as ARMAX and OE, will be briefly addressed in the second part of this section. An example completes the treatment of the Least Squares method.

5.3.1   Linear Least Squares

The problem that is addressed is the following:

Given measurements of input and output $\{u(k), y(k)\}_N$, and given a model structure with unknown parameters, determine the parameters of a model, that describes the process behavior as accurately as possible.

The notion "as accurately as possible" is of course highly subjective. Therefore the criterion, mentioned in Section 5.1, will be used (see (5.4)):

$$J_N(\theta) = \frac{1}{N} \sum_{k=1}^{N} \epsilon^2(k|\theta) \tag{5.40}$$

As was stated before, the estimated parameter vector, denoted by $\hat{\theta}_N$, is found to be

$$\hat{\theta}_N = \arg \min_{\theta \in D_M} J_N(\theta) \tag{5.41}$$

For FIR and ARX model structures the regression model is linear in θ:

$$\hat{y}(k|\theta) = \phi^T(k)\theta \tag{5.42}$$

The criterion (5.40) then becomes

$$J_N(\theta) = \frac{1}{N} \sum_{k=1}^{N} \left[y(k) - \phi^T(k)\theta \right]^2 \tag{5.43}$$

In the optimal solution $\hat{\theta}_N$, the gradient of $J_N(\theta)$ with respect to θ is zero:

$$\frac{\partial}{\partial \theta} J_N(\hat{\theta}_N) = 0 \tag{5.44}$$

From linear algebra it is known that

$$\frac{\partial}{\partial \theta} \left[y(k) - \phi^T(k)\theta \right]^2 = -2 \frac{\partial}{\partial \theta} \left[\phi^T(k)\theta \right] \cdot \left[y(k) - \phi^T(k)\theta \right]$$
$$= -2\phi(k) \left[y(k) - \phi^T(k)\theta \right]$$

Then (5.44) becomes:

$$\frac{\partial}{\partial \theta} J_N(\hat{\theta}_N) = -\frac{1}{N} \sum_{k=1}^{N} 2\phi(k) \left[y(k) - \phi^T(k)\theta \right] = 0$$

Hence the optimal solution $\hat{\theta}_N$ can be found by solving the following set of equations, also called the *Normal Equations*:

$$\left[\frac{1}{N} \sum_{k=1}^{N} \phi(k)\phi^T(k) \right] \hat{\theta}_N = \frac{1}{N} \sum_{k=1}^{N} \phi(k)y(k) \tag{5.45}$$

It is sometimes more convenient to have a matrix-vector notation for (5.45). This is obtained by defining the following matrix Φ and vector Y:

$$
\Phi = \begin{bmatrix} \phi^T(1) \\ \phi^T(2) \\ \vdots \\ \phi^T(N) \end{bmatrix} \qquad Y = \begin{pmatrix} y(1) \\ y(2) \\ \vdots \\ y(N) \end{pmatrix} \tag{5.46}
$$

The criterion function then becomes

$$
J_N(\theta) = |Y - \Phi\theta|_2^2 \tag{5.47}
$$

where $|.|_2$ represents the Euclidian norm. The Normal Equations (5.45) are written as

$$
\left[\Phi^T \Phi\right] \hat{\theta}_N = \Phi^T Y \tag{5.48}
$$

from which the solution $\hat{\theta}_N$ is obtained via

$$
\hat{\theta}_N = \left[\Phi^T \Phi\right]^{-1} \Phi^T Y \tag{5.49}
$$

In Section 8.5 a numerically robust way of solving the Least Squares problem is given.

5.3.2 Pseudo-Linear Least Squares

The derivation above was based on (5.42). The crucial part is the assumption that $\hat{y}(k|\theta)$ is linear in the parameter vector θ (FIR, ARX). If this is not the case, such as for ARMAX or OE prediction models, solving the Least Squares problem becomes more tedious.

In the pseudo-linear regression models, the regression vector depends on the parameter vector θ. The criterion function becomes

$$
J_N(\theta) = \frac{1}{N} \sum_{k=1}^{N} \left[y(k) - \phi^T(k|\theta)\theta\right]^2 \tag{5.50}
$$

and

$$
\hat{\theta}_N = \arg \min_{\theta \in D_M} J_N(\theta) \tag{5.51}
$$

To solve this pseudo-linear Least Squares problem a nonlinear iterative optimization technique must be used. These techniques are discussed in Chapter 8.

A drawback of these nonlinear methods is their iterative and consequently time-consuming character. Moreover, convergence cannot be guaranteed.

5.3.3 Example

The following example illustrates the use of the Least Squares method for parameter estimation.

Suppose that a real process is represented by the following equation:

$$
y(k) - 1.5y(k-1) + 0.7y(k-2) = u(k-1) + 0.5u(k-2) + e(k)
$$

The input $u(k)$ is ZMWN with $\sigma_u^2 = 1$, and the noise $e(k)$ is also ZMWN with $\sigma_e^2 = 1$. u and e are independent.

To identify a model for this process, a model representation is chosen. The selected model structure is ARX of second order:

$$y(k) = \phi^T(k)\theta = \left(\begin{matrix} -y(k-1) & -y(k-2) & u(k-1) & u(k-2) \end{matrix} \right) \begin{pmatrix} a_1 \\ a_2 \\ b_1 \\ b_2 \end{pmatrix}$$

To estimate the parameters of the ARX model, the Least Squares method is used. The Normal Equations (5.45) of the associated minimization problem become

$$\begin{bmatrix} \hat{R}_y(0) & \hat{R}_y(1) & -\hat{R}_{yu}(0) & -\hat{R}_{yu}(1) \\ \hat{R}_y(1) & \hat{R}_y(0) & -\hat{R}_{yu}(-1) & -\hat{R}_{yu}(0) \\ -\hat{R}_{yu}(0) & -\hat{R}_{yu}(-1) & \hat{R}_u(0) & \hat{R}_u(1) \\ -\hat{R}_{yu}(1) & -\hat{R}_{yu}(0) & \hat{R}_u(1) & \hat{R}_u(0) \end{bmatrix} \hat{\theta} = \begin{bmatrix} -\hat{R}_y(1) \\ -\hat{R}_y(2) \\ \hat{R}_{yu}(1) \\ \hat{R}_{yu}(2) \end{bmatrix}$$

where $\hat{R}$ demonstrates that the quantities must be estimated from a finite-length (N) data set.

A typical parameter estimate $(N = 200)$ is

$$\hat{\theta} = \left(\begin{matrix} -1.5207 & 0.7228 & 0.9883 & 0.4969 \end{matrix} \right)^T$$

which is close, but not equal to the true parameter vector $(-1.5, 0.7, 1, 0.5)^T$.

Asymptotically $(N \to \infty)$, the estimates of the covariance functions tend to the exact values. With the procedure outlined in Section 3.7 these exact values can be computed analytically, as

$$\begin{aligned} R_y(0) &= 27.7344 & R_{yu}(-1) &= 0 \\ R_y(1) &= 24.7656 & R_{yu}(0) &= 0 \\ R_y(2) &= 17.7344 & R_{yu}(1) &= 1 \\ & & R_{yu}(2) &= 2 \end{aligned} \tag{5.52}$$

Filling in these values, the asymptotic parameter estimate $\hat{\theta}$ is

$$\hat{\theta} = \left(\begin{matrix} -1.5 & 0.7 & 1 & 0.5 \end{matrix} \right)^T$$

We see that asymptotically $(N \to \infty)$ the original parameters have been found exactly, and the question arises whether this is always the case for an infinite data set. This problem is addressed in the next section.

5.4 Analysis of the linear LS estimate

The example in the previous section showed that asymptotically the original parameters are found exactly by means of the Least Squares method. Is this a coincidence, and if not, is it always the case, or is it only the case under special circumstances? That is the question considered in this section.

First an example will show that the original parameters are **not** always recovered.

Example 5.1 Consider the following process:

$$y(k) + a_0 y(k-1) = b_0 u(k-1) + e(k) + c_0 e(k-1) \tag{5.53}$$

The input and noise are both ZMWN, with covariances σ_u^2 and σ_e^2, respectively. The auto- and cross correlations can be computed analytically (for $N \to \infty$) as

$$R_y(0) = \frac{(1 + c_0^2 - 2a_0 c_0)\sigma_e^2 + b_0^2 \sigma_u^2}{1 - a_0^2}$$

$$R_y(1) = \frac{(c_0 - a_0 - a_0 c_0^2 + a_0^2 c_0)\sigma_e^2 - a_0 b_0^2 \sigma_u^2}{1 - a_0^2}$$

$$R_{yu}(0) = 0$$

$$R_{yu}(1) = b_0 \sigma_u^2$$

Suppose that we want to identify the following ARX-model:

$$y(k) + a y(k-1) = b u(k-1) + e(k) \tag{5.54}$$

with the Least Squares method. The Normal Equations then become

$$\begin{bmatrix} R_y(0) & -R_{yu}(0) \\ -R_{yu}(0) & R_u(0) \end{bmatrix} \begin{pmatrix} a \\ b \end{pmatrix} = \begin{pmatrix} -R_y(1) \\ R_{yu}(1) \end{pmatrix} \tag{5.55}$$

The solution then is

$$\hat{a}_\infty = a_0 - \frac{c_0 \sigma_e^2 (1 - a_0^2)}{(1 + c_0^2 - 2a_0 c_0)\sigma_e^2 + b_0^2 \sigma_u^2} \tag{5.56a}$$

$$\hat{b}_\infty = b_0 \tag{5.56b}$$

Obviously, the estimated $\hat{a}_\infty$ is not equal to the true a_0: there is a bias. The parameter $\hat{b}_\infty$ is exact. This is not coincidental, and will be explained later in this section. □

The bias in the parameter estimate can be explained by analyzing the LS estimate.
Suppose that the true system is given by

$$y(k) = \phi^T(k)\theta_0 + v_0(k) \tag{5.57}$$

The LS estimate is given by (5.49):

$$\hat{\theta}_N = \left[\Phi^T \Phi\right]^{-1} \Phi^T Y \tag{5.58}$$

Let the $d \times d$ covariance matrix of the regression vector be denoted by $\mathcal{R}_\phi^d$, where d is the number of parameters to be estimated ($d = \dim \theta$). Then (5.58) can be rewritten as

$$\hat{\theta}_N = [\mathcal{R}_\phi^d]^{-1} \frac{1}{N} \sum_{k=1}^{N} \phi(k) \left[\phi^T(k)\theta_0 + v_0(k)\right] \tag{5.59}$$

and, consequently

$$\hat{\theta}_N = \theta_0 + [\mathcal{R}_\phi^d]^{-1} \frac{1}{N} \sum_{k=1}^{N} \phi(k) v_0(k) \tag{5.60}$$

It is clear that the parameter estimate is unbiased if and only if the second part of the right hand side of (5.60) is zero. In Example 5.1 this was nonzero, leading to a biased parameter estimate. However, only the a-parameter was biased, since the covariance matrix of u was diagonal. In that case the equation for $\hat{b}$ becomes:

$$\hat{b} = R_{yu}(1)/\sigma_u^2 \qquad (5.61)$$

and clearly there is no influence of v_0 on this estimate. Therefore $\hat{b}$ in Example 5.1 is unbiased.

It is clear from (5.60) that the Least Squares Estimate exists if $\mathcal{R}_\phi^d$ is invertible. This requires the input sequence $u(k)$ to be persistently exciting of order d.

From this analysis we see that the LS Estimate is asymptotically unbiased if and only if the following two conditions are met:

1. $\mathcal{R}_\phi^d$ is invertible ($u(k)$ is persistently exciting of order n).

2. $\bar{\mathbb{E}}\{\phi(k)v_0(k)\} = 0$, which will be true in either of the following two cases:

   - $v_0(k)$ is ZMWN. In this case $v_0(k)$ does not depend on what happened up to time instant $k - 1$. Since $\phi(k)$ only contains samples up to time $k - 1$, the condition is met.
   - The following three conditions are satisfied:
     (a) the input sequence $u(k)$ is independent of $v_0(k)$
     (b) either $u(k)$ or $v_0(k)$ has zero mean
     (c) $n_a = 0$, which implies that $\phi(k)$ only contains samples of $u(k)$ (e.g., FIR)

There is a class of identification methods, called the Instrumental Variable (IV) methods, that explicitly uses (5.60). These methods require auxiliary signals, called *instruments*, that assure the second part in (5.60) to be zero. Instrumental Variable methods are treated extensively by Söderström and Stoica (1983).

5.5 *Convergence and consistency*

In the previous section we derived the conditions under which the LS estimate is asymptotically unbiased. This analysis can also be used for other (pseudo-linear) Prediction Error Methods. In this section the answer is given to the following two questions:

1. Does the parameter estimate converge to a certain value, as $N \to \infty$?

2. If the parameter estimate converges, does it converge to the true value θ_0?

The above properties are called *convergence* and *consistency*, respectively.

With a general PEM, the parameter estimate $\hat{\theta}_N$ is obtained according to (5.51):

$$\hat{\theta}_N = \arg \min_{\theta \in D_{\mathcal{M}}} J_N(\theta) \qquad (5.62)$$

where the loss function $J_N(\theta)$ is given by (5.50):

$$J_N(\theta) = \frac{1}{N} \sum_{k=1}^{N} \epsilon^2(k, \theta) \qquad (5.63)$$

As the number of samples N tends to infinity ($N \to \infty$), under weak conditions (Ljung, 1987) $J_N(\theta)$ tends to a limit function $J_\infty(\theta)$ with probability one, as stated in the following lemma.

Lemma 5.1 *Consider the model structures discussed in this chapter. The loss function (5.63) satisfies*

$$\sup_{\theta \in D_\mathcal{M}} |J_N(\theta) - J_\infty(\theta)| \to 0 \qquad w.p.\ 1 \qquad as\ N \to \infty \qquad (5.64)$$

with

$$J_\infty(\theta) = \lim_{N \to \infty} \frac{1}{N} \sum_{k=1}^{N} \epsilon^2(k, \theta) = \bar{\mathbb{E}} \left\{ \epsilon^2(k, \theta) \right\} \qquad (5.65)$$

But if the loss function $J_N(\theta)$ tends to the limit function $J_\infty(\theta)$, the minimizing argument $\hat{\theta}_N$ tends to θ_∞, where θ_∞ is the minimizing argument of $J_\infty(\theta)$. If there is more than one minimizing argument of $J_\infty(\theta)$, there is a set, denoted by D_∞:

$$D_\infty = \arg\min_{\theta \in D_\mathcal{M}} J_\infty(\theta) = \left\{ \theta | \theta \in D_\mathcal{M}, J_\infty(\theta) = \min_{\eta \in D_\mathcal{M}} J_\infty(\eta) \right\} \qquad (5.66)$$

Then we have the following convergence result.

Theorem 5.1 *Let $\hat{\theta}_N$ be defined by (5.62) and (5.63). Then, under weak conditions,*

$$\hat{\theta}_N \to D_\infty \qquad w.p.\ 1 \qquad as\ N \to \infty \qquad (5.67)$$

where D_∞ is given by (5.66).

Remark Convergence into a set as in Theorem 5.1 means that

$$\inf_{\theta_\infty \in D_\infty} \left| \hat{\theta}_N - \theta_\infty \right| \to 0 \qquad as\ N \to \infty \qquad (5.68)$$

∎

The above holds for a Prediction Error Method, applied to any of the model structures described in Chapter 3. We can now give a definition of a convergent estimator.

Definition 5.1 (Convergent estimator) *An estimator is called* convergent *if the parameter estimate $\hat{\theta}_N$ satisfies (5.67).*

As was said before, *consistency* means that the true parameter θ_0 can be retrieved. Therefore a consistent estimator can be defined as follows.

Definition 5.2 (Consistent estimator) *Suppose that the true parameter of a system is θ_0. An estimator is called* consistent *if the following two conditions both hold:*

1. *The estimator is convergent;*

2. *The set $D_\infty = \{\theta_0\}$*

Note that it is only possible to obtain a consistent parameter estimate, if the parameter vector can describe the system exactly. In this case "the system is in the model set".

If the system is in the model set, both $G(q)$ and $H(q)$ can tend to the real transfer functions $G_0(q)$ and $H_0(q)$, respectively. If the system is not in the model set, we are dealing with *approximate* identification. This is discussed in the next section.

The question whether $G_0(q)$ and $H_0(q)$ can be retrieved exactly from an identification experiment is often referred to as *identifiability*.

5.6 *Approximate identification*

Suppose that a Prediction Error Method is used for identification. The value θ_∞, to which the parameter estimate converges, is called the *limit value*. With this limit value a *limit model* is associated. Let us examine this limit model.

The limit loss function is

$$J_\infty(\theta) = \lim_{N \to \infty} \mathbb{E}\left\{\epsilon^2(k, \theta)\right\} \tag{5.69}$$

We then have with (3.42)

$$J_\infty(\theta) = \frac{1}{2\pi} \int_{-\pi}^{\pi} \Phi_\epsilon(\omega, \theta)\, d\omega \tag{5.70}$$

where $\Phi_\epsilon(\omega)$ is the spectrum of the prediction errors. From (5.70) an interesting relation can be derived.

Suppose that the real process is given by $G_0(q)$, and that the output can be written as

$$y(k) = G_0(q)u(k) + v_0(k) \tag{5.71}$$

where $v_0(k)$ represents any kind of noise, with arbitrary color, generated by filtering ZMWN with the filter $H_0(q)$.

The estimated transfer functions of process and noise process are denoted by $\hat{G}(q, \theta)$ and $\hat{H}(q, \theta)$, respectively.

The prediction error $\epsilon(k, \theta)$ satisfies

$$
\begin{aligned}
\epsilon(k, \theta) &= \hat{H}^{-1}(q, \theta)\left[y(k) - \hat{G}(q, \theta)u(k)\right] \\
&= \hat{H}^{-1}(q, \theta)\left[[G_0(q) - \hat{G}(q, \theta)]u(k) + v_0(k)\right]
\end{aligned} \tag{5.72}
$$

We then find for the spectrum $\Phi_\epsilon(\omega)$

$$\Phi_\epsilon(\omega) = \frac{\left|G_0(e^{i\omega}) - \hat{G}(e^{i\omega}, \theta)\right|^2 \Phi_u(\omega) + \Phi_v(\omega)}{\left|\hat{H}(e^{i\omega}, \theta)\right|^2} \tag{5.73}$$

provided that u and v_0 are independent.
The noise spectrum $\Phi_v(\omega)$ in (5.73) is given by

$$\Phi_v(\omega) = \left|H_0(e^{i\omega})\right|^2 \sigma_e^2 \tag{5.74}$$

The limit loss function $J_\infty(\theta)$ then is, using (5.70)

$$J_\infty(\theta) = \frac{1}{2\pi} \int_{-\pi}^{\pi} \frac{\left|G_0(e^{i\omega}) - \hat{G}(e^{i\omega}, \theta)\right|^2 \Phi_u(\omega) + \Phi_v(\omega)}{\left|\hat{H}(e^{i\omega}, \theta)\right|^2} \, d\omega \qquad (5.75)$$

A characterization of D_∞ in the frequency domain is then

$$D_\infty = \arg\min_\theta J_\infty(\theta) \qquad (5.76)$$

From (5.75)–(5.76) we can derive several important properties. First of all we see that if the system is in the model set, $\left|G_0(e^{i\omega}) - \hat{G}(e^{i\omega}, \theta)\right|$ can become zero (model → process), and a minimum of the loss function is obtained if the noise model tends to the real noise process. This is in accordance with our consistency results in the previous section.

What happens if the system is **not** in the model set? No theoretical results can be given, but we can analyze (5.75) in an intuitive way. We distinguish two different cases, depending on the parametrization of the noise:

1. fixed noise model

2. general case

These cases are discussed in the following two subsections.

The limit loss function (5.75) indicates an interesting approach, using filtered prediction errors, instead of the prediction errors themselves. In this way we can shape the model fit in the frequency domain. This approach is treated in Subsection 5.6.3, and illustrated by an example in Subsection 5.6.4.

5.6.1 Fixed noise model

For a fixed noise model $H(q, \theta) = H^*(q) \; \forall \theta$, (5.75)–(5.76) can be rewritten as

$$D_\infty = \arg\min_\theta \int_{-\pi}^{\pi} \left|G_0(e^{i\omega}) - \hat{G}(e^{i\omega}, \theta)\right|^2 Q^*(\omega) \, d\omega \qquad (5.77a)$$

$$Q^*(\omega) = \frac{\Phi_u(\omega)}{|H^*(e^{i\omega})|^2} \qquad (5.77b)$$

where we have disposed of θ-independent terms.

For any $\theta^* \in D_\infty$, $\hat{G}(e^{i\omega}, \theta^*)$ is a best mean-square approximation of $G_0(e^{i\omega})$, with a frequency weighting $Q^*(\omega)$. The frequency weighting $Q^*(\omega)$ depends on the input spectrum $\Phi_u(\omega)$ and the fixed noise model $H^*(q)$. The estimate $\hat{G}(e^{i\omega}, \theta^*)$ can thus be influenced by suitable choices of $\Phi_u(\omega)$ and $H^*(q)$.

A particularly interesting case occurs if an Output Error model is used. For OE models the (fixed) noise model is $H^*(q) = 1$. The parameters of the process model are then determined by (5.77a), where the frequency weighting is given as

$$Q^*(\omega) = \Phi_u(\omega) \qquad (5.78)$$

Hence the frequency weighting depends on the input signal only.

5.6.2   General noise model

If a general noise model is used, no exact characterization of the resulting parameter estimates can be given. In the special case where the noise model is independently parametrized (no common parameters in process and noise model, e.g., Box-Jenkins), Ljung (1987) derives an expression that shows that the noise model is constructed such that the noise model spectrum $|\hat{H}(e^{i\omega},\theta)|^2$ resembles the error spectrum as much as possible. The error spectrum is the spectrum of the error $y(k) - \hat{G}(q)u(k)$.

For other models, such as ARX and ARMAX, no theoretical results can be given. An intuitively clear observation, however, is that in general the resulting estimate θ^* is a compromise between fitting $\hat{G}(e^{i\omega},\theta)$ to $G_0(e^{i\omega})$ in the quadratic frequency norm

$$Q(\omega,\theta^*) = \frac{\Phi_u(\omega)}{|\hat{H}(e^{i\omega},\theta^*)|^2} \tag{5.79}$$

and fitting the noise model spectrum $|\hat{H}(e^{i\omega},\theta^*)|^2$ to the error spectrum.

5.6.3   Filtered prediction errors

It has already been shown in Example 5.1 that fitting an ARX-model to data that comes from a process, that cannot be described by an ARX-model, results in a biased parameter estimate. An advantage of ARX-models, however, is that the associated prediction error problem is linear in θ. Hence the solution can be computed analytically, which saves a lot of computation time compared to, e.g., pseudo-linear estimation problems. It would therefore be desirable to have the fast computation of linear least squares, but to have a better estimate. How can we improve the linear LS-estimate?

Let us first analyze the LLS estimate in the frequency domain. The process and noise models are given by (5.22). From the previous sections we can conclude that the asymptotic estimate θ_∞ is computed as

$$\theta_\infty = \arg\min_\theta$$

$$\int_{-\pi}^{\pi} \left\{ \left| G_0(e^{i\omega}) - \frac{B(e^{i\omega},\theta)}{A(e^{i\omega},\theta)} \right|^2 Q_{\text{ARX}}(\omega,\theta) + |A(e^{i\omega},\theta)|^2 \Phi_v(\omega) \right\} d\omega \tag{5.80}$$

where the frequency weighting $Q_{\text{ARX}}(\omega,\theta)$ is given by

$$Q_{\text{ARX}}(\omega,\theta) = |A(e^{i\omega},\theta)|^2 \Phi_u(\omega) \tag{5.81}$$

Concentrating on the estimate $\hat{G}(q,\theta)$, we see that it is fitted to $G_0(q)$ in the quadratic frequency norm (5.81). Since $1/A(q)$ is in general low-pass, its inverse is high-pass. This means that the frequency weighting is larger for higher frequencies. Hence the fit will be better for high frequencies than for low frequencies. This is confirmed by the following example.

Example 5.2 Consider the following system:

$$y(k) - 1.5y(k-1) + 0.7y(k-2) = u(k-1) + 0.5u(k-2)$$
$$+ e(k) - e(k-1) + 0.2e(k-2) \tag{5.82}$$

This system is clearly of the ARMAX type.

A white noise input $u(k)$ is applied, with variance $\sigma_u^2 = 1$. Data is collected for $N = 100$ samples.

A second order ARX-model is fitted to the data. In Figure 5.2 the Bode plots of the real system $G_0(q)$ (solid) and the resulting ARX-estimate $\hat{G}(q)$ (dashed) are shown.

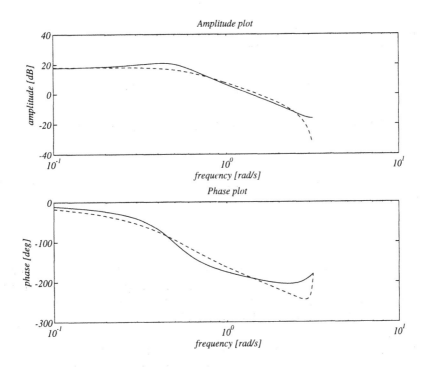

Figure 5.2 Bode plots of true system (solid) and ARX-estimate (dashed)

Obviously the estimate is better for high frequencies than for low frequencies. The dynamics of the system are hardly covered by the estimate. This is confirmed by the pole location of the process and the model, given in Table 5.1.

□

Table 5.1
Pole locations of the process and the ARX-estimate

System poles	ARX model poles
$0.7500 \pm 0.3708i$	$0.6230 \pm 0.2782i$

This effect is usually unwanted: the most interesting region is in general the low-frequency region.

To improve the estimate for low frequencies (at the cost of lower accuracy for higher frequencies), use can be made of *filters*. Instead of the prediction error

$\epsilon(k, \theta)$, we can use the filtered prediction error $\epsilon_F(k, \theta)$:

$$\epsilon_F(k, \theta) = L(q)\epsilon(k, \theta) \tag{5.83}$$

where $L(q)$ is a finite-dimensional, inversely stable, linear, time-invariant filter. The spectrum $\Phi_{\epsilon_F}(\omega)$ is given by

$$\Phi_{\epsilon_F}(\omega) = |L(e^{i\omega})|^2 \Phi_\epsilon(\omega) \tag{5.84}$$

where $\Phi_\epsilon(\omega)$ is given by (5.73).
The limit cost function $J_\infty(\theta)$ is then given by

$$J_\infty(\theta) = \frac{1}{2\pi} \int_{-\pi}^{\pi} |L(e^{i\omega})|^2 \Phi_\epsilon(\omega) \, d\omega \tag{5.85}$$

For the ARX-model the frequency weighting $Q_{\text{ARX}}(\omega)$ now becomes

$$Q_{\text{ARX}}(\omega) = |L(e^{i\omega})|^2 |A(e^{i\omega}, \theta)|^2 \Phi_u(\omega) \tag{5.86}$$

There are several ways to choose the low-pass filter $L(q)$. The quality of an OE-model is achieved if it is chosen as

$$L(q) = \frac{1}{A(q)} \tag{5.87}$$

However, this would require knowledge of $A(q)$. Since this is not available, a possible approach is to estimate $1/A(q)$ in a first step, and use the resulting estimate in the next step as the filter $L(q)$. This could be repeated a number of times, until the results are satisfactory. This approach is illustrated by an example in the next section.

The idea of using filtered prediction errors can be used for any model structure. It can be very useful to be able to shape the frequency weighting by an additional filter, especially if we cannot redesign an experiment.

Remark For SISO systems, filtering the prediction error with a filter $L(q)$ is equivalent to filtering input and output with that filter $L(q)$. Hence, instead of the data $\{u(k), y(k)\}_N$, we use the filtered data $\{u_F(k), y_F(k)\}_N$. ∎

5.6.4 Example

To illustrate the approach of using filtered prediction errors, we apply the procedure to the same system as in Example 5.2:

$$y(k) - 1.5y(k-1) + 0.7y(k-2) = u(k-1) + 0.5u(k-2) + e(k) - e(k-1) + 0.2e(k-2) \tag{5.88}$$

The same data ($N = 100$) is used as before. The estimated $\hat{A}(q)$ of Example 5.2 is used for the filter $L(q)$:

$$L(q) = \frac{1}{\hat{A}(q)} = \frac{1}{1 - 1.2459q^{-1} + 0.4655q^{-2}} \tag{5.89}$$

The data $\{u(k), y(k)\}_N$ is filtered with this $L(q)$, and then the Least-Squares method is applied to obtain a new second order ARX-model. The resulting

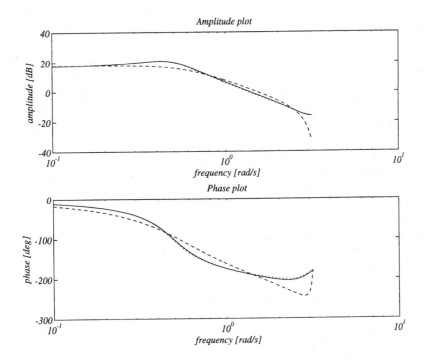

Figure 5.3 Bode plots of the true system (solid), the unfiltered ARX-estimate (dashed), and the ARX estimate with prefilter (dotted)

estimate is shown in Figure 5.3 (dotted), together with the true system (solid) and the first ARX-estimate (dashed) of Example 5.2.

It is clear that the quality of the estimate has improved considerably by prefiltering the data. This is confirmed by the pole locations of the true system, the ARX-model without, and the ARX-model with prefiltering, given in Table 5.2.

Table 5.2
Pole locations of the process, the unfiltered ARX-estimate
and the ARX-estimate with data prefiltering

System poles	ARX model poles	ARX model poles, filtered
$0.7500 \pm 0.3708i$	$0.6230 \pm 0.2782i$	$0.7591 \pm 0.3681i$

This example demonstrates the usefulness of filters for identification. They can be used to shape the resulting estimate according to the user's wishes.

5.7 *Summary*

This chapter has dealt with the problem of parametric identification. It has been shown that parametric models can be obtained by Prediction Error Methods.

These are based on the minimization of a criterion function, which in turn is based on the prediction error: the mismatch between measured output and model output.

One-step ahead prediction models have been derived for the FIR, ARX, ARMAX and OE model structures. The FIR and the ARX prediction models turned out to be linear in the parameter vector θ, which enables us to solve the minimization problem analytically. The ARMAX and OE prediction models are pseudo-linear, so the parameter vector must be found by means of some nonlinear optimization technique.

The Least Squares method was introduced to solve the optimization problem for linear model structures (FIR, ARX). Analysis of the resulting estimate showed that certain conditions have to be fulfilled, to obtain an unbiased estimate (Section 5.4).

After that, the concepts of convergence, consistency and identifiability were discussed. It was shown that Prediction Error Methods are convergent. Moreover, if the system can be described by the model set (system in the model set), PEMs were shown to be consistent.

Since in practice the system to be identified is never in the model set, we usually deal with approximate identification. This has been treated in Section 5.6. The discussion was based on a frequency-domain analysis of the limit model, which is the model that results if an infinite number of data is available. It was shown that in general the resulting estimate is a compromise between fitting the estimated $\hat{G}(e^{i\omega}, \theta)$ to $G_0(e^{i\omega})$ in a quadratic frequency norm, and fitting the noise model spectrum to the error spectrum.

Finally it was shown that by using prefilters, the estimates can be improved considerably.

5.8 *References*

Parameter estimation has been a research topic for years now. Many authors have done research in this area, resulting in a vast amount of literature on Maximum Likelihood estimators, Least Squares estimators, and realization theory, among which: Richalet *et al.* (1971), Graupe (1972), Eykhoff (1974), Mehra and Lainiotis (1976), Beck and Arnold (1977), Sorenson (1980), Sinha and Kuszta (1983), Caines (1988), Hannan and Deistler (1988), and Schoukens and Pintelon (1991).

The development and generalization of present-day Prediction Error Methods is ascribed to Ljung (1987). Similar work has been done by Söderström and Stoica (1988).

The asymptotic expressions and their implications for prefiltering, treated in Section 5.6, have first been derived by Wahlberg and Ljung (1986).

As pointed out in Section 5.4, the Instrumental Variable method was treated extensively by Söderström and Stoica (1983). A less thorough treatment can be found in (Ljung, 1987) and in (Söderström and Stoica, 1988).

For on-line identification we can use a *recursive* implementation of Least Squares, such as treated by Ljung and Söderström (1983), and various other authors.

Finally we mention the Matlab System Identification Toolbox (Ljung, 1991), in which several of the previously discussed identification methods are implemented. Given a data set of measured inputs and outputs, one can derive models with these functions.

5.9 *Problems*

1. Suppose that we have a process with dead time d. We want to model this with a first order ARX model:

$$y(k) + ay(k-1) = bu(k-d) + e(k)$$

   Derive the Normal Equations for problem of estimating a and b, on the basis of measurements of y and u.

2. Derive the prediction model for the BJ model structure, and write it in pseudo-linear regression form.
   To this end, determine the parameter vector θ and the regression vector $\phi(k|\theta)$.

chapter six

Identification in practice

In the previous two chapters we have discussed several methods of obtaining a model from a given data set. But how do we obtain such a data set? In general we design an experiment to extract data from a process. Typical ingredients of the design of experiments are the selection of inputs and outputs, the design of an input signal, the choice of the sampling frequency, and data (pre)processing. This problem is addressed in the first section of this chapter.

After collecting the data, we have to choose a model structure. Within the model structure, we have to select the order of the polynomials, and the dead time. This is discussed in Section 6.2.

After the experiment and the identification, the resultant model has to be evaluated. We want to have confidence in our model. In Section 6.3, several model validation techniques are discussed.

The contents of this chapter serve as a guideline to those who want to do real experiments for obtaining data sets for identification purposes.

6.1 Experiment design

If we are asked to make a black-box model of a process, we have to design an experiment. We have to choose the inputs and outputs, and we have to specify the way in which we are going to extract data from the process. Generally, these choices depend on the situation at hand, such as the availability of equipment to generate and measure the signals we want, and the dynamics of the process to be identified. There is no global theory on experiment design, since it is much too problem-dependent. However, there are some ideas that can be frequently applied.

The more *a priori* information about the process we have, the better the experiment we can design. Consequently, it is not surprising that experiment design is often conducted iteratively: preliminary experiments provide a preliminary model, on which the real experiment is based. Preliminary experiments could consist of a so-called *free-run* experiment, in which data is collected in normal operation, without exciting the process. Evaluation of the free-run data can provide information about the main dynamics of the process.

The experiment design variables are those variables that influence the result of the identification. Let us recall the asymptotic cost function J_∞ from the previous chapter:

$$J_\infty(\theta) = \frac{1}{2\pi} \int_{-\pi}^{\pi} |L|^2 \frac{|G_0 - \hat{G}(\theta)|^2 \Phi_u(\omega) + \Phi_v(\omega)}{|\hat{H}(\theta)|^2} \, d\omega \qquad (6.1)$$

in which $e^{i\omega}$ is omitted for readability. The limit model θ_∞ satisfies

$$\theta_\infty = \arg \min_\theta J_\infty(\theta) \qquad (6.2)$$

From (6.1) we see the dependency of θ_∞ on several variables. Although (6.1) only holds for an infinite number of data points, we assume that for finite N the variables have a similar effect. The design variables can be separated in two classes: variables that have to be chosen *a priori*, that is before the experiment is carried out, and variables that can be chosen *a posteriori*, that is when the data has been collected. The *a priori* design variables discussed here are the input spectrum $\Phi_u(\omega)$, the sampling frequency ω_s and an anti-aliasing filter. The *a posteriori* design variables are the model parametrization, the prefilter $L(q)$, and other data (pre)processing techniques.

6.1.1   Input design

Before the experiment is carried out, we have to design an input signal. The dependency of the parameter estimate $\hat{\theta}$ on the input signal is clear through the spectrum $\Phi_u(\omega)$ in (6.1).

It has already been pointed out in Section 5.6 that the estimated model will be closer to the true system in frequency regions where the input spectrum is large compared to the noise spectrum $\Phi_v(\omega)$. This is also intuitively clear, since the estimate at a certain frequency should become better as the amount of energy put into the system at that frequency becomes larger.

Another important factor is the degree of persistence of excitation of the input signal. This degree should be sufficiently high with respect to the number of parameters to be estimated. In this respect a Zero Mean White Noise signal is the ideal input signal, since it is persistently exciting of any order. However, apart from these considerations we are also faced with practical constraints. Although white noise is a nice input signal from a theoretical point of view, for most systems it is not allowed to apply such a signal, especially because of the high-frequency excitation. Moreover, since most processes act as low-pass filters, what is the use of putting a lot of energy in the high-frequency region, where there are almost no dynamics? In this case, other signals should be considered, such as a multisine (sum of sines), designed such that there is not much high-frequency excitation.

An input signal which is easy to design is Generalized Binary Noise (GBN), as presented by Tulleken (1990). A GBN sequence switches between two values (e.g., -1 and $+1$), with a nonswitching probability p ($0 < p < 1$). By varying p, the characteristics of the signal are influenced. Note that a GBN with $p = 1/2$ is a Pseudo Random Binary Sequence (PRBS). A larger p results in a signal with more energy in the lower frequencies.

6.1.2 Sampling frequency

Another choice, to be made before carrying out the experiment, concerns the sampling frequency ω_s. We distinguish between the sampling frequency for data collection, and the sampling frequency that is used in the identification procedure.

Data collection

For data collection we just want to sample as fast as possible. Choosing the sampling frequency too low will lead to loss of information, since the main dynamics of the process may then be too fast to be observed. A lower bound on the sampling frequency is obtained by applying Shannon's sampling theorem, which states that to recover a continuous-time signal exactly, the sampling frequency ω_s should be chosen such that the signal does not contain any frequencies above the Nyquist frequency $\omega_s/2$. In practice this cannot be guaranteed, but a suitable choice for the sampling frequency is:

$$\omega_s \geq 10 \cdot \omega_B \tag{6.3}$$

where ω_B is the bandwidth of the process. The bandwidth ω_B of the process is defined as the maximum frequency ω for which the magnitude of the frequency function reaches the level of $1/\sqrt{2}$ times its static value (-3dB), and can, for example, be determined from preliminary experiments. To avoid aliasing effects, an anti-aliasing filter has to be added to the system. An anti-aliasing filter is a low-pass filter, with cut-off frequency around the Nyquist frequency. Note that the filter can only be realized with continuous (analogue) components, before the sampling takes place!

Identification

For identification we have to take into account that choosing the sampling frequency too high will lead to numerical problems, since the poles of a discrete-time system are pushed towards the point $z = 1$ in the complex z-plane. This can be seen by looking at the relation between a pole z_p in the discrete-time domain and its equivalent s_p in the continuous-time domain:

$$z_p = e^{s_p T_s} = e^{2\pi s_p/\omega_s}$$

where T_s is the sampling time.

The higher the sampling frequency ω_s, the smaller T_s, and the more the pole is pushed towards $z_p = 1$. Another consideration is that we use one-step ahead prediction, and the smaller the sampling time, the smaller the prediction step. This implies that a high sampling frequency will emphasize the high-frequency fit of the model, which might not be what we want.

A practical choice of the sampling frequency is

$$10 \cdot \omega_B \leq \omega_s \leq 30 \cdot \omega_B \tag{6.4}$$

The sampling frequency for data acquisition should be higher than, or equal to, the sampling frequency for identification. The former is chosen as high as possible, the latter is chosen in accordance with the intended model application. To obtain the measured signals at a lower sampling frequency, they must be filtered again by an anti-aliasing filter! This filter can be implemented in software as a digital filter. The reduction of sampling frequency (an integer number) is called *decimation*. Note that the input signal must be designed on the basis of the sampling time for identification!

6.1.3   Data preprocessing

When the data is collected, the parameter estimate can still be influenced by the choice of the parametrization, the prefilter $L(q)$, and other possible data processing.

Data preprocessing is concerned with handling the data before estimating a model parameter. Possible goals are the decimation discussed previously, and the removal of outliers, spikes or trends in the data. An example of the latter is the removal of the mean of the signals, since in most identification methods we assume signals with zero mean. Another preprocessing operation is the scaling, to equalize their variance. This is especially useful for identifying MIMO systems.

In general data preprocessing is done in a computer. Several packages, such as Matlab, provide dedicated functions for trend removal, filtering and decimation. Removal of outliers and spikes is mostly done by visual inspection of the data.

6.1.4   Prediction error prefilter

The prediction error prefilter $L(q)$ strongly influences the identification results. In the previous chapter it was already shown that an ARX-estimate is improved considerably by applying a low-pass filter. Note in particular that the role of the filter $L(q)$ is equal to the role of the inverse of a fixed noise model $\hat{H}(q, \theta) \equiv H^*(q)$. The prefilter should be designed such that the frequency region of interest is emphasized.

Note that, as mentioned before, for SISO systems filtering the prediction error is equivalent to filtering the input and output, and can be seen as data preprocessing. However, this does not hold for MIMO systems, and therefore we rather view it as part of the identification procedure.

6.2 Model structure selection

With respect to the parametrization, we have already seen that an ARX-model is easy to estimate, since it has an analytical solution. However, the estimated parameter can be biased. More complex parametrizations (ARMAX) can result in better parameter estimates, but require the solution of nonlinear optimization problems. We also have the choice of using independently parametrized process and noise models (FIR, OE, BJ). No general guidelines can be given in this respect, except that the choice should depend on the intended model application. For example, if a model is needed for simulation, and we are not interested in the noise properties, it is obvious that an OE-parametrization suffices. We need not put effort in estimating a noise model. In other cases the choice of a model parametrization may not be so clear.

Once a specific parametrization is chosen, we have to select the model order, and we have to estimate the dead time of the process.

6.2.1 Determination of the model order

The model order should not be selected too low, since then not all dynamics can be described. It should not be selected too high either, since the higher the model order, the more parameters need to be estimated, and the higher the variance of the estimates is. Hence the model order is some sort of compromise. How can we choose it?

Having data available, we can compute the loss function $J_N(\hat{\theta}_N)$ (5.4) for models of different order. Theoretically, for $N \to \infty$, the loss function should tend to a minimum value for increasing model order. Increasing the model order beyond the true order of the process, will not add to the quality of the model. Hence we would expect a graph as shown by the solid line in Figure 6.1. The minimum of the loss function does not decrease for orders higher than five,

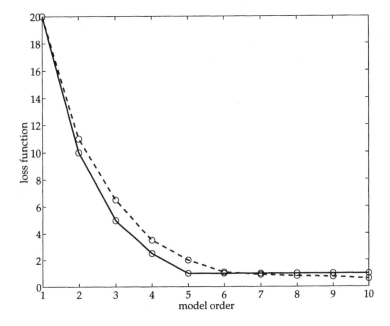

Figure 6.1 Loss function J_N for models of different order, theoretical (solid) and in practice (dashed)

and hence we conclude that the model order should be five.

In practice, however, there are several reasons why the curve will slowly decrease for increasing model order, as shown by the dashed curve in Figure 6.1. The main reason is that we can never expect the process to lie in the model set. Hence there is no such thing as a *true order* of the process.

A second reason is given by the finiteness of the data set. Since N is finite, the specific noise realization, present in the data, plays a role. If we select a very high order, it is possible to let the loss function go to zero. However, in this case we have not modeled the process dynamics, but the specific noise contribution in the data set. For a different data set, with a different noise realization, the model will not be adequate.

The effect of the noise realization can be avoided by calculating the value of J_N not with the data set, from which the model is estimated, but to use a different (part of the) data set. This is called *cross validation* (Stoica *et al.*, 1986). Using cross validation, the dashed curve in Figure 6.1 will become flatter again.

The model order selection procedure just described, is based on visual inspection of the loss function. It would be convenient to have an automated procedure. We want to find the smallest model order, with which the process can be described. Hence we can write the problems of model order selection and parameter estimation as a nested minimization problem, depending on the number of parameters d_θ to be estimated ($d_\theta = \dim \theta$), and the minimum loss function for each model order.

$$\{d_\theta, \hat{\theta}_N\} = \arg \min_{d \in \mathbb{N}} f\left(d, \min_{\theta \in D_\mathcal{M}} J_N(\theta)\right) \qquad (6.5)$$

where $f(.)$ is a mathematical function.
The inner minimization is the parameter estimation problem.

Several possibilities of choosing f in (6.5) can be found in literature. We mention two of them, proposed by Akaike (1969, 1974, 1981).

The first possibility is Akaike's Information Criterion (AIC).

$$d_\theta = \arg \min_{d \in \mathbb{N}} \log\left[\left(1 + \frac{2d}{N}\right) J_N(\hat{\theta}_N)\right] \qquad (6.6)$$

The second possibility is Akaike's Final Prediction Error (FPE).

$$d_\theta = \arg \min_{d \in \mathbb{N}} \frac{1 + d/N}{1 - d/N} J_N(\hat{\theta}_N) \qquad (6.7)$$

Both the AIC and the FPE are implemented in the Matlab System Identification Toolbox (Ljung, 1991).

6.2.2 Estimation of dead time

If there is dead time in a process, the process output does not react instantly to the applied input signal. The time period between application of the input and reaction of the process is called *dead time*, denoted by d.
Dead time is measured in the number of samples, and therefore it is a positive integer: $d \in \mathbb{Z}^+$. We assume all processes to be strictly proper, and hence they have a dead time of at least one.

A dead time d means that the first d parameters of the impulse response are zero. An example of a model with dead time d is the following first order ARMAX model.

$$y(k) + ay(k-1) = bu(k-d) + e(k) + ce(k-1)$$

In black-box identification, dead time can be incorporated in two ways. Either the first d coefficients of the B-polynomial are set to zero, or the input sequence is shifted over d samples. But then d must be estimated accurately.

What happens if we have estimated d incorrectly?

If we have underestimated the dead time, too few parameters of the B-polynomial are set to zero, and the identification procedure will usually result in several parameters that are almost zero. Dynamically, this often results in a nonminimum-phase behavior of the model, even though the process is minimum phase.

Underestimation of the dead time can be determined by inspection of the parameter estimates. If the first parameters of the B-polynomial are almost zero, it is likely that they should be set to zero, thus increasing the dead time estimate.

If we have overestimated the dead time, too many parameters of the B-polynomial are set to zero. In this case the identification method must compensate this incorrectness with the other parameters. In general this results in biased estimates.

Concluding, we state that overestimating the dead time is worse than underestimating.

How do we estimate the dead time? As we have pointed out, dead time is characterized by a delayed reaction of the output to an input signal. This might suggest a lack of correlation between the input and the output for the first $d - 1$ lags.

Dead time can be estimated by using correlation analysis, which has been discussed in Section 4.1. Recalling (4.3), we see that the cross-covariance function of y and u reflects the impulse response of the process.

$$R_{yu}(\tau) = \sum_{\ell=1}^{\infty} g_0(\ell) R_u(\tau - \ell) \tag{6.8}$$

where we have assumed that u and e are independent.
We see that if u is white noise with variance σ_u^2, this becomes

$$R_{yu}(\tau) = g_0(\tau)\sigma_u^2 \tag{6.9}$$

Consequently, if the dead time is d, $R_{yu}(\tau)$ is zero for $\tau \leq d - 1$.

From this we can conclude that the dead time of a process can be estimated by applying a white noise signal to the input, and determining the cross-covariance function of y and u. A hypothesis test (see also the next section) can be used to determine whether or not $R_{yu}(\tau)$ should be considered zero for several values of τ.

In practice one will not use white noise as an input signal. In that case prewhitening filters are used for estimating the dead time. Suppose that $H_u(q)$ is a whitening filter for u, such that the signal $u_H(k) = H_u(q)u(k)$ is white with variance σ^2. Furthermore, let $y_H(k) = H_u(q)y(k)$ be the output signal, filtered through the same whitening filter. Then we can calculate the cross-covariance function of y_H and u_H.

$$R_{y_H u_H}(\tau) = \mathbb{E}\left\{ H_u(q) \sum_{\ell=1}^{\infty} g_0(\ell)u(k - \ell) \cdot u_H(k - \tau) \right\}$$

$$= \sum_{\ell=1}^{\infty} g_0(\ell) R_{u_H}(\tau - \ell)$$

$$= g_0(\tau)\sigma^2 \tag{6.10}$$

where again we have used the fact tha u and e are uncorrelated. As in (6.9), the cross-covariance function will be zero for the first $d-1$ lags, if the process has a dead time of d samples.

Note that the above procedure can also be used for estimating the impulse response of a process. The estimate will be rough, since the cross-covariance function is estimated from a finite data set.

Of course one can also identify several models with different dead times, and analyze the resulting models to determine which dead time fits best.

6.3 Model validation

One of the problems in black-box identification is that we always obtain a model, whether it makes sense or not. Therefore we need a means to evaluate the correctness, the validity, of the model. We have to put a tag on the model, indicating our confidence in it. Without this tag, the model is worthless. If validation indicates that the model is not good, some of the design variables of the experiment should be changed, and the identification redone. Sometimes even the experiment has to be redone.

How can we attach a quality tag to a model? It would be nice if we could *prove* the correctness of a model, but this is only possible if we have the real process available (in which case we do not need identification). It follows that we can only try to show that the model is invalid. If we do not manage to invalidate the model, we can assume it to be correct.

In this light the term *model validation* does not really represent what we do; we should rather call it *model invalidation*. In this section several ways of invalidating a model are discussed. Note that this is not an exhaustive list. Any method you can think of is a good way to invalidate a model.

In general a distinction is made between *objective* and *subjective* invalidation techniques. An objective method is based on mathematical properties of model and data set, and the resulting conclusion does not depend on the user. A subjective method is largely influenced by the user: different users can come up with different conclusions. The first method to be discussed is an objective method, followed by some subjective techniques. In practice one should apply several invalidation techniques, since different methods reveal different errors.

Note that we want a model not only to explain the data from which it is built, but we want it to explain other data as well. Therefore, in general, the data set is split in two parts, to provide an identification set and a validation set. The model is estimated from the first data set, and the validation is based on the latter data set. We thus reduce effects that depend on the specific noise realization during the experiment.

6.3.1 Residual analysis

One way to evaluate a model is to look at the *residuals*. The residuals are those parts of the data that are not explained by the model. For a nonrecursive Prediction Error Method, the residual equals the prediction error $\epsilon(k,\theta)$:

$$\epsilon(k,\theta) = \hat{H}^{-1}(q,\theta)[y(k) - \hat{G}(q,\theta)u(k)] \qquad (6.11)$$

We see that if the model is correct, that is $\hat{G}(q, \hat{\theta}) = G_0(q)$ and $\hat{H}(q, \hat{\theta}) = H_0(q)$, the residual will tend to a white noise sequence $e(k)$:

$$\epsilon(k, \theta_0) = e(k) \tag{6.12}$$

where θ_0 is the real parameter value.

If the parameter is estimated correctly, $\epsilon(k, \hat{\theta})$ should have two properties, that $e(k)$ has:

1. $e(k)$ is white noise, and hence the autocovariance function has a specific form;

2. $e(k)$ is independent of $u(k)$, which implies that their cross-covariance function is zero.

An intuitive interpretation of the second property is that if the residual is independent of the input, all parts of the input have been "explained" by the model $\hat{G}(q)$. What is left over (the residuals) bears no additional information to improve $\hat{G}$.

Residual analysis now comes down to investigating the autocovariance function of the residuals, and the cross-covariance function of input and residuals. For both covariance functions we can develop hypothesis tests to decide whether or not the model should be accepted.

Whiteness of the residuals

To test the whiteness of the residuals ϵ, we proceed as follows. Without giving a proof we state that if ϵ is white, then

$$\sqrt{N}\frac{R_\epsilon(\tau)}{R_\epsilon(0)} \in As\mathcal{N}(0, 1)$$

where $As\mathcal{N}(0, 1)$ means that the given quantity asymptotically converges in distribution to the Normal distribution, with mean 0 and variance 1 (see, e.g., Ljung (1987)).

Defining N_α as the α-level of the $\mathcal{N}(0, 1)$ distribution, with

$$Pr\left\{ \left| \frac{R_\epsilon(\tau)}{R_\epsilon(0)} \right| \leq \frac{N_\alpha}{\sqrt{N}} \right\} = \alpha \tag{6.13}$$

we can define the following null hypothesis:

$$H_0 : \left| \frac{R_\epsilon(\tau)}{R_\epsilon(0)} \right| \leq \frac{N_\alpha}{\sqrt{N}} \tag{6.14}$$

We then test the null hypothesis, and if H_0 is accepted, the model is accepted. Note that if the number of samples N increases, the confidence limits (6.14) become tighter.

Remark α is the risk of invalidating a valid model (the probability that the test quantity lies outside the bounds, while in fact the model is good). Typical values for α are 0.01 and 0.05, resulting in confidence intervals of 99% and 95%, respectively. The associated confidence levels are $N_{0.01} = 2.58$ and $N_{0.05} = 1.96$. The risk that an invalid model is not rejected cannot, in general, be computed. Of course, if α decreases, this risk increases. ∎

To use this test for model validation, we compute the residuals, calculate the normalized autocovariance function as in (6.14), and plot them together with the confidence levels in one figure. If the normalized covariance function lies within the confidence region, we accept the null hypothesis (6.14) and hence the model. If the covariance function lies outside the confidence region, even if this happens for only one value of τ, we reject H_0, and the model is invalid. Note that we choose the confidence levels N_α ourselves, by specifying the probability of accepting an invalid model.

This procedure will be illustrated later by an example.

Remark The whiteness test is only meaningful if a noise model is estimated, otherwise (FIR, OE) we are almost sure to have nonwhite residuals.   ∎

Independence of residuals and input

Whiteness of the residuals indicates that both $\hat{G}$ and $\hat{H}$ are estimated correctly. However, often we are more interested in the correct estimation of the process model $\hat{G}$, and the nonwhiteness of the residuals does not necessarily mean that the model $\hat{G}(q)$ is incorrect. In that case we could use the independence of input and residuals and interpret it as follows: if there is correlation between the input and the residuals, there is still information in the residuals that could be put into the model. If the input and the residuals are independent, all information in the residuals is explained by the process model $\hat{G}$, and hence it is estimated correctly. Consequently, testing the independence of input and residuals provides a basis for the decision whether or not to accept the model $\hat{G}(q)$.

To test the independence of u and ϵ, we proceed as follows. If u and ϵ are indeed independent, we have that

$$\sqrt{N} R_{\epsilon u}(\tau) \in As\mathcal{N}(0, P) \tag{6.15}$$

with

$$P = \sum_{k=-\infty}^{\infty} R_\epsilon(k) R_u(k) \tag{6.16}$$

Using the same notation as before, the null hypothesis that the input and the residuals are independent becomes

$$H_0 : |R_{\epsilon u}(\tau)| \leq \sqrt{P/N} \cdot N_\alpha \tag{6.17}$$

In general, normalized quantities are used, and hence the null hypothesis is accepted if

$$\left| \frac{R_{\epsilon u}(\tau)}{\sqrt{R_\epsilon(0) R_u(0)}} \right| \leq \sqrt{\frac{P}{N R_\epsilon(0) R_u(0)}} \cdot N_\alpha \tag{6.18}$$

To evaluate a model, the normalized cross-covariance function of input and residuals is plotted, together with the confidence levels. In this way it can easily be seen whether or not (6.18) is satisfied. The model is accepted if (6.18) holds.

Note that, as in (6.14), if the number of samples N increases, the confidence limits (6.18) become tighter.

Remark Correlation between $\epsilon(k)$ and $u(k - \tau)$ for negative τ indicates the presence of feedback. It does not necessarily mean that the model structure is not sufficient. ∎

Remark If there appears to be too much correlation between $\epsilon(k)$ and $u(k - \tau)$ for positive τ, we would reject the model $\hat{G}$. However, we can also identify a (nonparametric) model between u and ϵ. Analyzing this model we might conclude that the dependency lies in "harmless" frequency regions, and still decide to accept the model $\hat{G}$ as it is. ∎

Example

An example illustrates the use of residual analysis for model validation. Consider a process, given by

$$[1 - 1.5q^{-1} + 0.7q^{-2}]y(k) = [q^{-1} + 0.5q^{-2}]u(k) + [1 - q^{-1} + 0.2q^{-2}]e(k) \quad (6.19)$$

We generate a data set of 4000 samples, and divide it into an identification set and a validation set, both containing 2000 samples. We identify (1) a second order ARMAX model, (2) a second order OE model and (3) a first order ARX model. We expect the ARMAX model to be correct, for the OE model we expect $\hat{G}$ to be correct, and we expect the ARX model to be incorrect. In Figure 6.2 the results from residual analysis are shown, where the validation set is used.

The upper plots show the autocovariance function of the residuals, the lower plots show the cross-covariance function of input and residuals. The dotted lines are the 99% confidence limits, calculated by Matlab.

From the plots in Figure 6.2a we conclude that the residuals of the ARMAX model are white (upper plot) and that they are independent of the input (lower plot). Consequently we accept both process model $\hat{G}(q)$ and noise model $\hat{H}(q)$, as expected.

From the plots in Figure 6.2b we see that the residuals of the OE model are not white (upper plot) since the normalized autocovariance function lies partly outside the confidence region. However, the residuals are independent of the input (lower plot), and hence we conclude that the process model $\hat{G}(q)$ is correct, and that the noise model $\hat{H}(q) \equiv 1$ is incorrect, as expected.

From the plots in Figure 6.2c we see that neither are the residuals of the ARX model white, nor are they independent of the input. Both the autocovariance function (upper plot) and the cross-covariance function (lower plot) lie partly outside the confidence interval, and hence the model is invalid, as expected.

Note that residual analysis also provides an idea how to adapt the model parametrization, to obtain better results. In this case one would conclude from Figure 6.2c that the order of the model is not high enough.

6.3.2 Other validation techniques

Most other model (in)validation techniques are subjective: depending on the user, the model is claimed to be valid or not.

The most straightforward way is to compare real and simulated outputs. If they differ too much, we conclude that the model is invalid. This type of validation is called *face validation*. Note that *differing too much* is a rather subjective criterion.

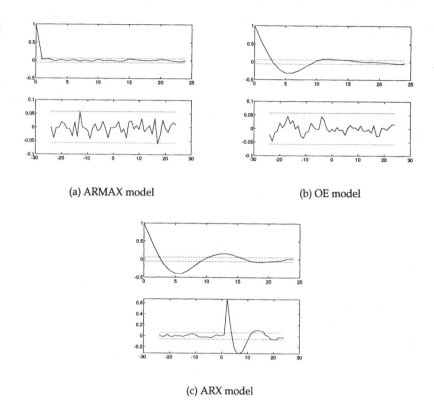

(a) ARMAX model   (b) OE model

(c) ARX model

Figure 6.2 Autocovariance function of residuals (upper plots) and cross-covariance function of input and residuals (lower plots) with 99% confidence intervals

Any other method on which a decision can be based whether or not to accept a model, is a possible validation technique. One can think of asking an expert if he can distinguish between real and simulated output, or one can compare the simulated output with data obtained from another model, which is usually a very high order, nonlinear white-box model.

The general advice is to do your best to show that a model is incorrect. If you cannot show this, your model can be accepted.

6.4 *Summary*

In this chapter we have discussed several practical aspects of identification. The estimated model depends on the input signal, the sampling time, the applied filters, and possible data preprocessing. These items have been discussed, and some guidelines with respect to their design have been given.

The selection of the model structure depends on the intended model application. Within a model structure, we have to select the order of the polynomials, and the dead time.

The model order can be selected by evaluating the loss function J_N for different model orders. Possible procedures that penalize the model complexity are Akaike's Information Criterion (AIC) and Akaike's Final Prediction Error (FPE). Using one of these selection procedures, the model order is a compromise between fitting the data and the model complexity.

The dead time can be estimated using time-domain correlation analysis. It should not be overestimated, because we would set too many model parameters to zero. Underestimating the dead time is less of a problem. By inspecting the model, it can be detected, and hence corrected.

In the third part of the chapter we discussed model validation, and stated that most validation techniques try to invalidate a model. An often used objective validation technique is residual analysis, which is based on the asymptotic result that the residuals should be independent of the input, and that they should be white, if the model is correct. Other, subjective, techniques can and should be used when trying to (in)validate a model.

6.5 References

Most of the literature on experiment design has focused on optimal input design. As pointed out by Mehra (1974, 1981), optimal input design is mainly based on an *information matrix*, representing the amount of information that is present when applying a certain input signal. Optimal input design is then concerned with the maximization, over the input signal, of the amount of information in the data. This can be achieved by maximizing some function of the information matrix, such as the determinant, the trace, or the ratio of the smallest and the largest eigenvalues. More information can be found in the papers already mentioned, and in contributions of Schroeder (1970), Federov (1972), Mehra and Lainiotis (1976), Goodwin and Payne (1977), Zarrop (1979), Sinha and Kuszta (1983), Tulleken (1990), Richalet (1991), and of Schoukens and Pintelon (1991).

The whole problem of experiment design, including input design, data preprocessing, and model structure selection, is discussed by Ljung (1987), and by Söderström and Stoica (1988). The latter have also included a chapter on closed-loop identification, in which they deal with the problem of estimating a model from closed-loop data. This situation often occurs in practice, when production and environment specifications require that a controller is present during the data collection. The problem then is, that the input signal becomes correlated with the noise, and hence special precautions must be taken to handle the data correctly.

The model validation problem has been addressed by Sinha and Kuszta (1983), Ljung (1987), Sargent (1988) and by Söderström and Stoica (1988).

6.6 Problems

1. Suppose that we identify an Output Error model of a process. We know that the model order is lower than the actual order of the process.
   We require the model to be good in the low-frequency region. This must be achieved by properly choosing the input signal u. Should we choose u mainly low-frequent, mainly high-frequent, or white noise?

2. Explain why a fixed noise model can act as a prefilter.

3. Suppose that we have data available from a process, excited with a white noise input. We want to identify an Output Error model. We know that the model order is lower than the actual order of the process.

   We require the model to be good in the low-frequency region. This must be achieved with a prefilter.

   (a) Should we choose the prefilter to be low-pass, band-pass or high-pass?

   (b) If we would use a fixed noise model, instead of a prefilter, should the noise model then be low-pass, band-pass or high-pass?

chapter seven

Simulation

7.1 Introduction

This chapter deals with the simulation of continuous and discrete mathematical models with the aid of the digital computer. Simulation is a tool for obtaining responses of these (nonlinear) models to analyze and understand their dynamic behavior. If models are linear, many methods are available for calculating the time responses, for example the inverse Laplace transformation and the Z-transformation, which transforms a continuous model into a discrete one after which the time responses can be calculated quite easily. When nonlinearities come into focus or when a model consists of both continuous and discrete parts, the techniques intended for linear models are no longer suitable. Other techniques have to be used.

We shall show why and how simulation can be applied and shall also indicate its limitations. Moreover, the limitations of simulation will be illustrated. These limitations relate to the compromise that can be and has to be made between accuracy and calculation time and the limited validity of the results of a simulation experiment. If these limitations are recognized, then simulation with the digital computer is a powerful and flexible tool.

Simulation is, among other reasons, used:

- If *no analytical solution* can be obtained of a complex model. By solving the numerical equations of the models with the aid of simulation techniques, an impression can be obtained of the dynamic behavior of this model. If simulation is not available, we have to resort to linearization, and this introduces errors.

- If *experiments with a real system* are:

    - too expensive (e.g., satellites in orbit),

    - too dangerous (e.g., nuclear power reactor),

    - impossible (e.g., society, human life, economic systems),

    - too fast or too slow (e.g., physical phenomena in the nucleus of atoms, evolution models);

- For *extrapolation* of measured data into the future (weather forecasting);

- For *sensitivity analysis* and parameter optimization studies of new processes.

Simulation is one of the most widespread tools for solving technical problems. It poses almost no restrictions on the description of the mathematical model. As long as a model can be described, its simulation, that is calculating the response of the model to some input or disturbance signal, is nearly always possible. Simulation is in use, for example, for:

- weather forecasting;

- computer and video games;

- explaining the past of earth, the climate, life;

- analyzing the economy;

- for deriving suitable models;

- for parameter estimation;

- for controller design.

It will be shown that simulation of both discrete and continuous models using the digital computer will introduce the following problems, namely:

- In a digital computer variables are represented by means of numbers. These numbers have a finite accuracy. This finite accuracy sometimes introduces problems;

- A model is assumed to represent a parallel system. Digital computers have only one or a finite number of processors. Consequently, the parallel-described simulation model has to be calculated with a sequential-oriented computer. This conversion from parallel to sequential can introduce problems;

- Continuous models are described by differential equations. It will be found that these differential equations can only be solved approximately. This approximation depends to a high degree on the selection of the integration methods and the size of the integration interval. We shall indicate how these quantities are to be selected.

The first two problems are common to for both discrete and continuous models. The last problem applies specifically to continuous models only.

Although simulation poses almost no restrictions on the way in which a model can be formulated, a distinction can still be made among the different mathematical models of dynamic systems (see also Section 2.3):

Continuous models

- Distributed-parameter models (partial differential equations)
- Lumped-parameter models (differential equations)
- State-event models with state-dependent events

Discrete models

- Sampled-data models (difference equations)
- Discrete-event models with time-dependent events

Each of these mathematical models can be used in appropriate simulation programs. The ordinary differential equation and difference equation are, in general, supported by a simulation program. In contrast, partial differential equations are hardly ever allowed. Only some programs yield facilities for describing a model with both differential or difference equations and events, arising from either state- or discrete-event processes.

7.2 Simulation tools

7.2.1 Simulation hardware

Nowadays, there is no debate whether an analog or a digital computer is most suited for the simulation of continuous models. After the long domination of analog computers, the fast developments in digital electronics have increased the importance of digital computers in the simulation of both discrete and continuous models. Only in very dedicated applications, where speed is of utmost importance, are analog computers still used. Still, a small description of the analog computer is valuable.

Analog computer

The analog computer consists of a number of standard elements such as amplifiers, variable gains, voltage sources, etc., which can be connected relatively easily to each other by wires. Based on these electronic elements, certain functions are realized. In Figure 7.1 a gain and an integrator are illustrated, realized with the aid of an operational amplifier and several passive elements, such resistors and capacitors.

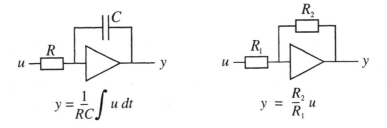

$$y = \frac{1}{RC}\int u\, dt \qquad\qquad y = \frac{R_2}{R_1} u$$

Figure 7.1 Gain and integrator

Having a number of these elements available, a user can connect them in a required way. The connected elements then form an electronic circuit which represents a continuous model. This short description of the analog computer at once reveals a number of its characteristics.

- Because the analog computer acts as a configurable electronic circuit, all calculations of all variables take place at the same time. This type of computer performs all calculations in parallel. As a consequence, the speed of calculation is high and does not depend on the size of the model. Only the bandwidth of the electronic components limits the speed of calculation. It is also possible to scale the time. By replacing all capacitors of the analog integrators by capacitors with one-tenth of the original capacity, the speed of calculation will increase with a factor 10. Of course, other reduction or magnifying factors can be utilized for decreasing or increasing calculation time.

- The variables are represented by electronic voltages and currents. Depending on the electronic components selected, these voltages have a maximum value (e.g., 150 [V]). Also a minimum value exists owing to the noise of the electronic components. This noise may require that a minimum value has to be larger than, for example, 1 [mV]. These upper and lower levels for the voltage pose a severe restriction on the size of all variables. In general, the variables of the mathematical model have to be scaled into an adjusted model such that the adjusted variables fit within these lower and upper levels. Too large values have to be avoided to prevent overflow and too small values have to be circumvented to maintain accuracy in the presence of noise. Consequently, scaling is a necessity. All variables have to be investigated for their smallest and largest values and scaled accordingly. Scaling is a difficult and tedious job which requires quite a lot of skill of the programmer.

- Each function or element of a model has to be available. If a given computer supports, for example, 10 integrators and the model requires 15, then it is impossible to implement such a model on that particular analog computer.

- Due to their limited applicability, the prices of analog computers are high and the availability of analog computers is poor.

- Analog computers are only suited for continuous models. The simulation of discrete elements such as difference equations, dead times, etc. is not possible by analog computer.

In view of these characteristics, we can state that analog computers offer a high calculating speed, and they are therefore they are still attractive for dedicated simulation purposes. However, their disadvantages, such as the necessity of scaling, the finite number of functions in a computer and the relatively high prices have made the digital computer preferable for executing simulations.

Digital computer

Since the arrival of the microcomputer and the ever increasing functionality and decreasing prices of digital electronics, the digital computer has provided ever increasing versatility in the performance of many different tasks, among them simulation.

For simulation purposes, the important characteristics of the digital computer are as follows.

- Any digital computer has one, or a finite number of calculating elements. This implies that all calculations have to be executed in series. Parallel execution of calculations is impossible. Consequently, measures have to be taken to avoid errors when parallel-defined models are calculated. Calculation time is approximately proportional to the size of a model.

- The variables in a digital computer are represented by numbers. These numbers have a limited accuracy and a limited range. If floating-point numbers are used in single precision accuracy, represented by 16 bits, the accuracy is about 6 digits (relative accuracy 10^{-6}) and a dynamic range from 10^{-36} to 10^{+36}. This large dynamic range and the relatively high accuracy make scaling superfluous.

- A digital computer is not able to calculate continuous variables. These variables have to be approximated at discrete time intervals. Consequently, continuous variables will be approximated by a sequence of points at discrete points in time. Also a continuous integrator requires an adequate substitute in the digital computer.

- As a consequence of their general applicability, digital computers offer a high performance for a relatively low price. Especially, the arrival of powerful PCs and workstations has provided an attractive simulation environment with good graphical facilities.

- A model with both continuous and discrete elements can be dealt with easily.

There are many advantages in using the digital computer for simulation, but there is one important drawback, namely calculation time. Here the analog computer is almost unbeatable. However, if we accept a finite calculation time, the digital computer is a very valuable simulation tool.

Accuracy

Although a relative accuracy of single-precision numbers in digital computers of 10^{-6} seems reasonable, numerical errors still occur as a consequence of roundoff processes. In particular, the subtraction of large numbers introduces a loss of accuracy. The following example illustrates the introduction of numerical errors.

Suppose the numbers a and b are represented in the digital computer with a relative accuracy δa and δb:

$$\delta a = \frac{\Delta a}{a} \qquad \delta b = \frac{\Delta b}{b} \tag{7.1}$$

Then the four basic arithmetic operators $*$, $/$, $+$ and $-$ introduce the following relative accuracy δc in c:

$$
\begin{aligned}
c &= a * b & \delta c &= \delta a + \delta b \\
c &= a / b & \delta c &= \delta a + \delta b \\
c &= a + b & \delta c &= \frac{\Delta a + \Delta b}{a + b} \\
c &= a - b & \delta c &= \frac{\Delta a + \Delta b}{a - b}
\end{aligned}
\tag{7.2}
$$

The relative accuracy δc of c after applying the operator $*$, $/$ or $+$ is about the relative accuracy of both a and b. The relative accuracy δc after subtraction can increase considerably. With $a = 1000$ and $b = 999$ it follows with $c = a - b$ that $\delta c = 10^3 * \delta a$. Consequently, the relative error is increased and the accuracy is decreased with a factor of 1000!

In Figure 7.2 the effect of the loss of accuracy owing to numerical manipulations is illustrated. The functions $f_i(x)$ are shown by

$$f_1(x) = (x-4)^6 \tag{7.3a}$$
$$f_2(x) = x(x(x(x(x(x-24)+240)-1280)+3840)-6144)+4096 \tag{7.3b}$$
$$f_3(x) = x^6 - 24x^5 + 240x^4 - 1280x^3 + 3840x^2 - 6144x + 4096 \tag{7.3c}$$

With x = 4, the values of $f_i(4)$ are calculated as

$$f_1(4) = (x-4)^6 = 0 \tag{7.4a}$$
$$f_2(4) = 4(4(4(4(4(-20)+240)-1280)+3840)-6144)+4096 = 0 \tag{7.4b}$$
$$f_3(4) = 4096 - 24576 + 61440 - 81920 + 61440 - 24576 + 4096 = 0 \tag{7.4c}$$

Mathematically, these functions $f_i(x)$ are equivalent. Numerically, they behave differently, as shown in Figure 7.2. All calculations are executed with single precision (about 6 to 7 digit accuracy). As can be observed the calculation of $f_3(x)$ introduces the largest errors, because in calculating $f_3(x)$ large numbers are subtracted.

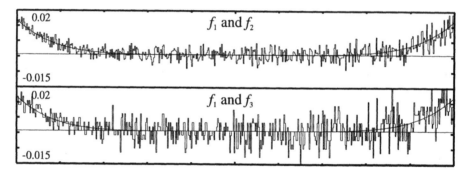

Figure 7.2 Numerical behavior $f_1(x)$, $f_2(x)$ and $f_3(x)$ for $x \in [3.5, 4.5]$

7.2.2 Simulation software

In this section some types of simulation software for the digital computer will be discussed. Instead of using this ready-made software, the user's own programming activities can be utilized to obtain the required solution of the simulation model. In general, such an approach is too expensive in man power because all facilities such as interaction, graphics, data storage, integration algorithms, etc. have to be devised. Such an approach can only be justified if dedicated requirements have to be fulfilled.

Simulation software can be divided into simulation programs and simulation languages.

Simulation programs

These programs have been designed since 1955 to emulate an analog computer on a digital one. So, all functions of the analog computer such as gains, integrators, function generators, adders, subtractors, multipliers are available and many more. Nowadays elements such as dead times, transfer functions, noise generators, trigonometric and logic functions, and all kinds of nonlinearities are supported. The connection between the elements, which are called blocks, are fixed by defining the inputs of each block. Subsequently, the parameters of all blocks are defined. Besides this structure and these parameters, the timing and output data have to be given. The timing data concern the integration method, integration interval T and the final time T_f of the simulation run. Although all variables of a simulation will be calculated, only a limited number of them can be shown, either as graphical responses on a screen or plotter or as numbers in a table on a printer. The output data define how the selected responses will be shown.

These four groups of data suffice to define both the simulation model and the simulation environment. A Run command will start the required simulation and show the indicated responses.

Simulation programs are block-oriented and interpretative. Both structure and parameters are stored in arrays. At any moment this information can be inserted, modified or used. No compilation step is required. Consequently, these programs can be highly interactive with immediate reaction of the program to the wishes of the user. Well-known simulation programs are Matrix$_x$, PSI/c, SIMNON, SIMULINK and TUTSIM.

Example 7.1 PSI/c has been developed at the Control Laboratory of the Delft University of Technology. The simulation model is a second order continuous model, namely

$$\ddot{y}(t) + \dot{y}(t) + y(t) = u(t) \tag{7.5}$$

Before this equation can be inserted into a simulation program or into a simulation language, it has to be rewritten into a number of first order integral equations and one algebraic equation, by means of:

$$\dot{y}(t) = \dot{y}(0) + \int_0^t \ddot{y}(\tau)\, d\tau$$

$$y(t) = y(0) + \int_0^t \dot{y}(\tau)\, d\tau \tag{7.6}$$

$$\ddot{y}(t) = u(t) - \dot{y}(t) - y(t)$$

These new equations can be represented in a block diagram which is suited for simulation. Each block receives a unique user-defined name. This name defines the output of the block as a variable. The block diagram is shown in Figure 7.3. The input $u(t)$ is assumed to be a unit step.

In PSI/c, this simple model is described in the following way.

```
% Dynamic specifications
Output    = INT(Deriv par: Output0);
    %y(t) is called Output
Deriv     = INT(U - Deriv - Output par: Deriv0);
    % dy/dt(t) is called Deriv
% Parameters
U         = 1;
Output0   = 0;        % Initial  value  y(0)
Deriv0    = 0;        % Initial  value  dy/dt(0)
```

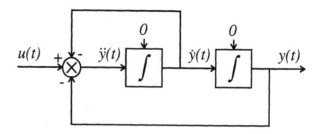

Figure 7.3 Block diagram of a second order model

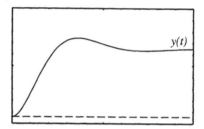

Figure 7.4 Response of $y(t)$, calculated with PSI

On the Run command the solution of $y(t)$ is calculated and immediately shown on the screen as illustrated in Figure 7.4.

Due to their attractive user-interface and the many simulation facilities, simulation programs are attractive as simulation tools. Their basic limitation is that no additional block types can be inserted or that the simulation model cannot be extended with statements of a higher programming language. Moreover, these programs deal with blocks with only one output. This restriction avoids the use of vectors and matrices, but this can be quite annoying when dealing with more complex models. However, new simulation programs yield additional facilities such as user-programmable blocks or the use of vectors. For models with up to a few hundred blocks simulation programs are very valuable.

Simulation languages

Simulation languages allow the description of models by means of equations instead of by single blocks. Still, continuous models have to be decomposed into single integrators. The other relations can be expressed by means of equations. In the simulation language CSMP III the model in Figure 7.4 becomes

```
INITIAL
    PARAMETER U=1.

DYNAMIC
    DERIV = INTGRL(0.,DERIV2)
    Y = INTGRL(0.,DERIV)
    DERIV2 = U - DERIV - Y

TIMER DELT = 0.1, FINTIM = 10.
OUTPUT Y
STOP
```

END

This model looks like a program. In fact it is written in the language as defined by CSMP III. If responses of this model are required, CSMP III translates this program into a Fortran program. With the aid of a regular Fortran compiler an object module and with a linker, a run module can be created. These activities are illustrated for the simulation language ACSL in the sequence of Table 7.1.

Table 7.1
ACSL simulation program

Status	Action	Adding
Fortran statements		
	Translation	- ACSL Macro's - User-defined Macro's
Fortran program		
	Compilation	
Object program		
	Linking	- ACSL libraries - Fortran libraries
Run module		
	Run time commands	
Responses		

Modifications to the structure require a complete new sequence of translation, compilation and linking and such modifications take quite a lot of time. Other modifications, for example the selection of other values for the parameters, the integration interval, the selection of the output variables can be achieved with the run-time commands. This yields a much quicker reaction.

Simulation languages offer additional flexibility since it is possible to include, for example, Fortran statements in the model description. So, nearly all types of models can be programmed. Still, it requires a higher skill to use simulation languages than to use simulation programs. ACSL is a well-known simulation language.

7.3 *Numerical solution of differential equations*

7.3.1 Introduction

There are several ways of obtaining a discrete model from a continuous model in order to calculate time responses; we may mention the following:

- Z-transformation

- Transforming the state equations

- Substitution of s by a function of z

- Numerical integration.

The first three approaches will be discussed briefly in order to illustrate their capabilities and limitations, and afterwards a more thorough treatment of numerical integration methods will follow.

Z-transformation

If the Laplace function is known of both the input $U(s)$ and the model $H(s)$, the output is given by $Y(s) = H(s) * U(s)$. With the aid of the Z-transformation, $Y(z)$ and, subsequently, $y(k)$ can be calculated, namely

$$Y(z) = Z\{\text{Hold}(s).H(s).U(s)\} \tag{7.7}$$

A hold circuit (e.g., zero order or first order) has been added to ensure that the value of $y(k)$ is maintained in between two samples. At the sample instants, the value of $y(k)$ corresponds exactly to the value of $y(t)$. So, the Z-transformation yields an accurate description of the continuous model. However, two requirements have to be fulfilled, namely

- the model has to be linear,

- the Laplace function of the input has to be known.

If one of these requirements is not fulfilled, additional hold circuits have to be introduced, for example to disconnect the input from the model and to isolate a possible nonlinearity from the linear parts, as illustrated in Figure 7.5.

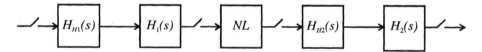

Figure 7.5 Introduction of hold circuits to allow the Z-transformation

The introduction of additional hold circuits introduces errors. Consequently, the output $y(k)$ no longer represents the continuous variable $y(t)$ accurately. A second drawback of using the Z-transformation is its lack of flexibility. If just one variable is changed in a transfer function $H(s)$, a complete new Z-transformation of a linear part has to be executed, because all parameters of the resulting $H(z)$ may be changed. For general simulation purposes the Z-transformation is not applied.

Transforming the state equation

A linear model can be described by

$$\dot{x}(t) = A.x(t) + B.u(t) \tag{7.8}$$

This continuous model can be transformed into

$$x(k+1) = P.x(k) + Q(k) \tag{7.9}$$

with

$$P = e^{AT} = I + AT + \frac{(AT)^2}{2!} + \frac{(AT)^3}{3!} + \cdots \qquad (7.10)$$

$$Q = \int_0^T e^{A(T-\tau)} Bu(\tau)d\tau$$

If A is invertible, and if the input $u(t)$ is constant between $t = kT$ and $t = (k+1)T$ with value $u(k)$, then the calculation of Q can be simplified, namely

$$Q = A^{-1}(e^{AT} - I)Bu(k) \qquad (7.11)$$

The calculation of P via the series expansion of e^{AT} converges, at least mathematically. In practice, due to roundoff problems, it requires the application of dedicated algorithms to achieve accurate results.

Still, this approach allows the calculation of responses by means of a discrete model of a continuous model. Accurate answers can be expected if the model is linear and the input u is constant between two samples. In other situations errors are introduced. Moreover, flexibility is poor because a complete recalculation of both P and Q is necessary if one element of matrix A changes its value. This approach is not suited for general simulation purposes.

Substitution of s by a function of z

Mathematically, the relation between s and z is defined by:

$$z = e^{sT} \quad \text{or} \quad s = \frac{1}{T}\ln(z) \qquad (7.12)$$

This nonlinear relation is not attractive for the substitution of s in an arbitrary polynomial of s, say $p(s)$. A more usable approximation of s by z is proposed by Tustin, by introducing the bilinear W-transformation $z = \frac{1+w}{1-w}$ and a Taylor approximation of $\ln(1 + x)$:

$$s = \frac{1}{T}\ln(z) = \frac{1}{T}\ln\left(\frac{1+w}{1-w}\right) = \frac{1}{T}(\ln(1+w) - \ln(1-w)) \simeq \frac{2w}{T} \qquad (7.13)$$

So,

$$\frac{1}{s} = \frac{T}{2} \cdot \frac{(1+z^{-1})}{(1-z^{-1})} \quad \text{or} \quad s = \frac{2}{T} \cdot \frac{(1-z^{-1})}{(1+z^{-1})} \qquad (7.14)$$

The continuous transfer function $H(s)$

$$H(s) = \frac{b_n s^n + b_{n-1}s^{n-1} + \cdots + b_0}{s^n + a_{n-1}s^{n-1} + \cdots + a_0} \qquad (7.15)$$

can now be rewritten into a discrete transfer function $H(z)$ by using the Tustin approximation a number of times for all powers of s in $H(s)$.

The Tustin substitution is an approximation. Consequently, the calculated $H(z)$ will not be exact and the results of calculating $y(k)$ on the basis of this $H(z)$ will yield some errors. Still, this substitution is a relatively flexible way of obtaining a discrete model from a continuous one.

This Tustin substitution can also be used if a discrete model $H(z)$ with a sample time T_1 has to be reformulated for another sample time T_2. Then, first use the inverse Tustin formula with T_1 to obtain $H(s)$ and then apply the direct Tustin formula with the new sample time T_2.

7.3.2   Numerical integration

Digital computers are not able to calculate exactly either a continuous differentiator or a continuous integrator. Still, a sufficiently accurate solution of first
order integral equations can be obtained with numerical integration methods.
Therefore, high order differential or integral equations have to be transformed
into a set of first order integral equations. Several approaches can be followed
in order to derive first order integral equations. Two of them will be dealt with
here.

Suppose a model is described by a transfer function H(s)

$$H(s) = \frac{b_n s^n + b_{n-1} s^{n-1} + \cdots + b_0}{s^n + a_{n-1} s^{n-1} + \cdots + a_0} \tag{7.16}$$

or by the differential equation

$$y^{(n)}(t) + a_{n-1} y^{(n-1)}(t) + \cdots + a_0 y(t) = b_n u^{(n)}(t) + \cdots + b_0 u(t) \tag{7.17}$$

This linear n^{th} order differential equation can be transformed into a set of n
linear first order differential equations by calculating an appropriate state space
model. Many different models can be selected, for instance the series or first
canonical form, the phase-variable or second canonical form and the parallel or
Jordan form. For illustrative purposes the first and second canonical forms will
be given.

The first canonical form is defined by

$$\dot{x}_i = b_{n-i} u - a_{n-i} y + x_{i+1} \tag{7.18a}$$

$$y = b_n u + x_1 \tag{7.18b}$$

and illustrated in Figure 7.6.

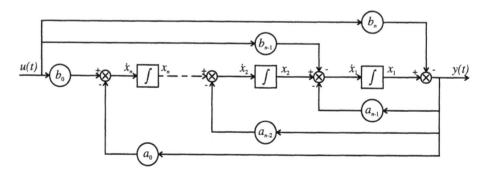

Figure 7.6 First canonical form of a differential equation

The states of the second or phase-variable form are selected as phase variables, so

$$\dot{x}_1 = x_2 \tag{7.19}$$

$$\vdots$$

$$\dot{x}_{n-1} = x_n$$

$$\dot{x}_n = u - a_0 x_1 - a_1 x_2 - \cdots - a_{n-1} x_n$$
$$y = (b_0 - a_0 b_n)x_1 + \cdots + (b_{n-1} - a_{n-1}b_n)x_n + b_n u$$
$$= (b_0 x_1 + b_1 x_2 + \cdots + b_{n-1}x_n) + b_n(u - a_0 x_1 - \cdots - a_{n-1}x_n)$$

These relations are illustrated in Figure 7.7.

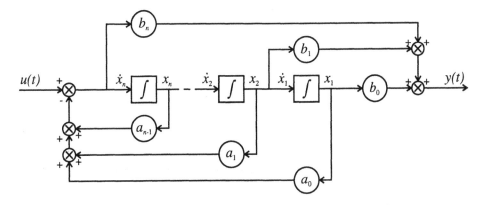

Figure 7.7 Second or phase-variable form of a differential equation

Based on these and other breakdowns of higher order differential equations or transfer functions a set of first order differential and, subsequently, integral equations can be made available. Also, nonlinear differential equations yield no additional problems. The equation

$$\ddot{y}(t) + \dot{y}(t) + y^2(t) = u(t) \tag{7.20}$$

can yield the integral equations

$$\dot{y}(t) = \dot{y}(0) + \int_0^t \ddot{y}(\tau)d\tau$$

$$y(t) = y(0) + \int_0^t \dot{y}(\tau)d\tau \tag{7.21}$$

$$\ddot{y}(t) = u(t) - \dot{y}(t) - y^2(t)$$

Now we can concentrate on the solution of first order integral equations, namely the solution of

$$y(t) = y(0) + \int_0^t f\left(y(\tau), \tau\right) d\tau \tag{7.22}$$

with $f(y(t), t)$ the input of the integrator. Due to the mathematical relation describing an integrator, $f(y(t), t)$ can be considered as the time derivative of $y(t)$. A model can be divided into two parts, one part containing all integrators, represented by the vector y, and the remaining part defining all other blocks or equations used to calculate the inputs $f(t)$ of the integrators, represented by the vector f. In any simulation program this distinction has to be made to solve the associated integral equations. This distinction is shown in Figure 7.8.

Just one solution of $y(t) = f(y, t)$ with $y(t_0) = y_0$ exists in a required area R of (t, y) if

- f is continuous in R

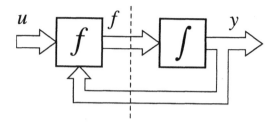

Figure 7.8 Dichotomy of a model into y and f

- for all points (t, y_1) and (t, y_2) the Lipschitz condition

$$|f(y_1, t) - f(y_2, t)| < L \, |y_1 - y_2| \qquad (7.23)$$

is satisfied. In practice this condition is met if $f(y, t)$ has a finite derivative with respect to $y(t)$.

This proposed first order integral equation represents an initial-value problem. So, the integral has to be solved starting with a fixed value $y(0)$ of the state for $t = 0$. There also exist two-point boundary value problems. These problems have fixed values for the state for $t = 0$ but also for $t = T_f$, the final time of the response. In general, a criterion is added to a two-point boundary value problem to select the, in some sense, optimal response that starts in $y(0)$ and ends in $y(T_f)$. The solution of this type of problems is difficult and very time-consuming.

Due to the character of a digital computer, continuous variables are approximated and sampled both in amplitude and in time. The approximation in amplitude is rather accurate (e.g., relative accuracy of 10^{-6}), the approximation in the time axis is rougher. Only at specific points in time the value of all variables will be calculated. The distance between two points in time is called the integration interval T, as illustrated in Figure 7.9.

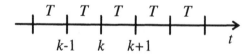

Figure 7.9 Division of time axis into periods

Suppose the simulation has succeeded until $t = kT$. Then the values of all variables up to and including t, and hence for $t = \ldots, (k-2).T, (k-1).T, k.T$, are known. Now the value of the output $y(t)$ of an integrator has to be approximated for $t = (k + 1).T$. This approximation can only be based on values of $y(t)$ and $f(t)$ up to and including $t = k.T$. For clarity, the points in time $y(kT)$ will be abbreviated as $y(k)$. A general integration method then becomes

$$y(k + 1) = F\{f(k + 1), f(k), f(k - 1), \ldots, y(k), y(k - 1), y(k - 2), \ldots\} \quad (7.24)$$

Depending on the shape of F, different integration methods are distinguished. By means of a number of definitions these different methods are characterized as follows.

Explicit and implicit methods: An integration method is called *explicit* if in function F in (7.24) the value of $f(k+1)$ is not required. If in a method the value of $f(k+1)$ is required for calculating $y(k+1)$, then this method is called implicit. Implicit methods give rise to problems if the input $f(k)$ depends on the value of the output $y(k)$.

One-step and multi-step methods: A *one-step* integration method is only based on some or all values of the input and output of *one* integration interval, so $f(k)$, $f(k+1)$ and $y(k)$. A *multi-step* method takes also the values of f and y outside this interval into account, for example $f(k-1)$ and/or $f(k-2)$. Multi-step methods are not self-starting because at $t = 0$ they lack knowledge of $f(-1), f(-2), \ldots$.

Order: The *order* of an integration method is determined by the highest degree of the polynomial $y(t)$ which can be represented accurately by $y(k)$. Suppose that $f(k) = c$ has a constant value. The output of the integrator $y(t)$ satisfies $y(t) = y(0) + c.t$, which is a first order polynomial in t. A first order integration method will yield accurate results. The same integration method will introduce errors if $f(t) = a + b.t$ because $y(t)$ will become a second order polynomial: $y(t) = y(0) + a.t + \frac{1}{2}b.t^2$. The higher the order of an integration method, the more accurate the results that can be achieved.

There are two means of obtaining suitable expressions for the function F, namely methods based on a Taylor-series approximation of $y(k)$ and methods based on a polynomial approximation of $f(k)$.

Taylor approximation of y(k)

The Taylor approximation of $y(k+1) = y(kT + T)$ is:

$$y(k+1) = y(k) + T.\dot{y}(k) + \frac{T^2}{2}.\ddot{y}(k) + O(T^3) \qquad (7.25)$$

If we neglect the derivatives of second and higher order and knowing that the input of an integrator is the derivative of its output: $\dot{y}(k) = f(k)$, the first integration method is constructed, namely the so-called *Euler* integration method:

$$y(k+1) = y(k) + T.f(k) \qquad (7.26)$$

The Euler method is a first order, one-step and explicit integration formula. In each integration interval 1 calculation of $f(k)$ is necessary. In simulation programs, the majority of calculation time is spent in calculating $f(k)$. Consequently, Euler will be a fast method.

Taking also the second order derivative of $y(k)$ into account will yield a second order integration method. However, the value of $\ddot{y}(k) = \dot{f}(k)$ is in general not available.

The *Runge-Kutta* integration methods have implemented an approach which replaces the necessity of knowing high order derivatives of $f(k)$. In the next example a second order Runge-Kutta integration method will be derived. It is assumed that

$$y(k+1) = y(k) + T.f(k) + \frac{T^2}{2}.\dot{f}(k) + O(T^3) \qquad (7.27)$$

The derivative will be approximated via a function ϕ, so

$$y(k+1) = y(k) + T.\phi \tag{7.28}$$

with ϕ a weighted sum of the derivatives ξ_1 and ξ_2 at k and $k + pT$, respectively, with $0 \leq p \leq 1$:

$$\phi = a.\xi_1 + b.\xi_2 = a.f(k) + b.f(k+p) \tag{7.29}$$

A Taylor expansion of ϕ offers:

$$\phi = a.f(k) + b.f(k) + bpT.\dot{f}(k) + O(T^2) \tag{7.30}$$

Combining yields

$$\begin{aligned} y(k+1) &= y(k) + T.\{a.f(k) + b.f(k) + bpT.\dot{f}(k) + O(T^2)\} \\ &= y(k) + (a+b).T.f(k) + bpT^2.\dot{f}(k) + O(T^3) \end{aligned} \tag{7.31}$$

Comparison with a second order Taylor expansion of $y(k+1)$ offers the following requirements in the selection of a, b and p, such that the derivative ϕ yields the same accuracy as taking also the first derivative of $f(k)$ into account:

$$\begin{aligned} a+b &= 1 \\ bp &= \tfrac{1}{2} \end{aligned} \tag{7.32}$$

This derivation of a second order Runge-Kutta method illustrates that many different combinations of a, b and p yield the accuracy of a second order integration method. For, if $y(t)$ is a second order polynomial of the time t, then $O(T^3) = 0$ and consequently, $y(k+1)$ yields the accurate value of $y(t)$. The following choices for a, b and p are often made:

$$\begin{array}{lll} a = 0 & b = 1 & p = \tfrac{1}{2} \\ a = \tfrac{1}{2} & b = \tfrac{1}{2} & p = 1 \end{array}$$

If we select $a=0$, $b=1$ and $p = \tfrac{1}{2}$, then the following equations result:

$$\xi_1 = f(k, y(k)) \tag{7.33a}$$

$$y^p\left(k + \frac{1}{2}\right) = y(k) + \xi_1 \frac{T}{2} \tag{7.33b}$$

$$\xi_2 = f\left(k + \frac{1}{2}, y^p\left(k + \frac{1}{2}\right)\right) \tag{7.33c}$$

$$y(k+1) = y(k) + \xi_2 T \tag{7.33d}$$

These relations can be interpreted as:
Determine with the aid of Euler the value of $y^p(k + \tfrac{1}{2}) = y(k) + \tfrac{1}{2}\xi_1 T = y(k) + \tfrac{1}{2}T.f(k)$. Based on this predicted value of the output, the predicted value of the input $f^p(k + \tfrac{1}{2})$ is calculated, with $\xi_2 = f^p(k + \tfrac{1}{2})$ at $t = (k + \tfrac{1}{2})T$. Only this value of the derivative is used for calculating the real value of $y(k + 1)$, after which the value of the input $f(k + 1)$ has to be calculated for the next integration step. So, in each integration interval two values of the input $f(k)$ have to be evaluated, namely $f(k)$ and $f^p(k + \tfrac{1}{2})$

Based on the formulas of Runge-Kutta in (7.33), it can be stated that they describe a second order, one-step, self-starting and explicit integration method.

Also higher order integration methods following the concept of Runge-Kutta are available. Especially, the *fourth order Runge-Kutta* method has been implemented in many simulation programs. The basic equations of the most-used implementation are:

$$\xi_1 = f(k, y(k)) \tag{7.34a}$$

$$\xi_2 = f\left(k + \frac{1}{2}, y(k) + \xi_1 \frac{T}{2}\right) \tag{7.34b}$$

$$\xi_3 = f\left(k + \frac{1}{2}, y(k) + \xi_2 \frac{T}{2}\right) \tag{7.34c}$$

$$\xi_4 = f\left(k + 1, y(k) + \xi_3 \frac{T}{2}\right) \tag{7.34d}$$

$$y(k+1) = y(k) + \frac{T}{6}\left(\xi_1 + 2.\xi_2 + 2.\xi_3 + \xi_4\right) \tag{7.34e}$$

This fourth order Runge-Kutta integration method requires four calculations of the input $f(k)$, namely at the points in time k, $k + \frac{1}{2}$, $k + \frac{1}{2}$ and $k + 1$. Compared with Runge-Kutta second order this requires approximately two times and compared with Euler four times more calculation time.

Remark Be aware that in general $\xi_2 \neq \xi_3$, although both approximate the value of the derivative $f(k + \frac{1}{2})$! ∎

Polynomial approximation of $f(k)$

A second way to calculate an appropriate function F in (7.24) for numerical integration methods is to use polynomial inter- or extrapolation. Based on old values of $f(k)$, such as $f(k-1)$, $f(k-2), \ldots$, a polynomial $p(t)$ is analytically calculated which can be used to define a continuous approximation for $f(t)$ with $kT \leq t \leq (k+1)T$. We shall illustrate this idea by a number of examples of low order polynomial approximations.

Suppose $f(t)$ is approximated by a zero order polynomial $p_0(t)$ which coincides with $f(k)$ at $t = k$, as illustrated in Figure 7.10.

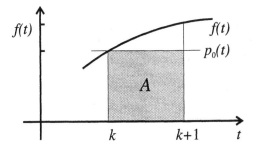

Figure 7.10 Zero order approximation of $f(t)$

The area A below the polynomial $p_0(t)$ is given by

$$A = T.f(k) \tag{7.35}$$

The contribution to the integral from $t = kT$ to $t = (k+1)T$ is A, so that

$$y(k+1) = y(k) + A = y(k) + T.f(k) \tag{7.36}$$

This equation turns out to be the Euler integration method. In the same way second order integration methods can be derived by using a first order polynomial $p_1(t)$ for approximating $f(t)$. In Figure 7.11 two different methods are introduced for determining $p_1(t)$, the first based on the values of $f(k-1)$ and $f(k)$ and the second based on knowledge of $f(k)$ and $f(k+1)$.

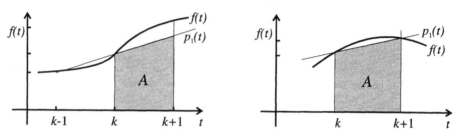

Figure 7.11 Second order integration methods

In Figure 7.11a the polynomial $p_1(t)$ is determined by the values of $f(k-1)$ and $f(k)$. The area A below $p_1(t)$ for $kT \leq t \leq (k+1)T$ is

$$A = T. \left\{ f(k) + \frac{1}{2}(f(k) - f(k-1)) \right\} = \frac{T}{2}\{3.f(k) - f(k-1)\} \qquad (7.37)$$

yielding the so-called *second order Adams-Bashforth* integration method

$$y(k+1) = y(k) + \frac{T}{2}\{3.f(k) - f(k-1)\} \qquad (7.38)$$

In Figure 7.11b the polynomial $p_1(t)$ is determined by $f(k)$ and $f(k+1)$, offering an area A with

$$A = \frac{T}{2}\{f(k) + f(k+1)\} \qquad (7.39)$$

The associated integration method, which is called the *second order Adams-Moulton* method, looks like

$$y(k+1) = y(k) + \frac{T}{2}\{f(k) + f(k+1)\} \qquad (7.40)$$

It can be noted that higher order polynomials $p_i(t)$ can also be selected to approximate the integral over $f(t)$. Analytical integration of these polynomials leads to two different classes of formulas, namely the formulas according to Adams-Bashforth and the formulas according to Adams-Moulton. The difference is located in the presence of the value of $f(k+1)$ in the integration formula. With this term they belong to the Adams-Moulton methods and without this term to the Adams-Bashforth integration methods. These methods are illustrated in Table 7.2.

Apart from the first order methods, the higher order methods of both Adams-Bashforth and Adams-Moulton are multi-step integration methods. Because the value of $f(k+1)$ is not used in the Adams-Bashforth formulas, these integration methods are explicit. In contrast, the formulas according to Adams-Moulton are implicit as a consequence of the presence of $f(k+1)$.

Another characteristic of both types of formula is that they are not *self-starting*. For $t = 0$, the values of $f(k-1)$, $f(k-2)$, $\cdots$ are not known. Errors are introduced if these values are assumed to be zero. Before these methods can be

Table 7.2
Integration methods based on polynomial approximation

Order	Adams-Bashforth
1	$y(k+1) = y(k) + Tf(k)$
2	$y(k+1) = y(k) + \frac{T}{2}\{3f(k) - f(k-1)\}$
3	$y(k+1) = y(k) + \frac{T}{12}\{23f(k) - 16f(k-1) + 5f(k-2)\}$
4	$y(k+1) = y(k) + \frac{T}{24}\{55f(k) - 59f(k-1) + 37f(k-2) - 9f(k-3)\}$
Order	Adams-Moulton
1	$y(k+1) = y(k) + Tf(k+1)$
2	$y(k+1) = y(k) + \frac{T}{2}\{f(k+1) + f(k)\}$
3	$y(k+1) = y(k) + \frac{T}{12}\{5f(k+1) + 8f(k) - f(k-1)\}$
4	$y(k+1) = y(k) + \frac{T}{24}\{9f(k+1) + 19f(k) - 5f(k-1) + f(k-2)\}$

applied, accurate values of the derivative $f(k)$ have to be calculated, for example by using Runge-Kutta methods of the same order as that of the chosen Adams-Bashforth or Adams-Moulton method.

In each integration interval just one calculation of the derivative $f(k)$ has to be made, even for higher order formulas. This indicates that the *multi-step* integration formulas are relatively fast compared with Runge-Kutta higher order methods.

As can be expected intuitively, the implicit Adams-Moulton formulae give a better performance compared than the corresponding Adams-Bashforth formulae. This improved performance can be attributed to the availability of knowledge of the derivative $f(k+1)$ at $t = (k+1)T$. However, this improved performance is to some extent offset by the disadvantage that these methods cannot be applied to models where the input of an integrator depends on the value of the output of the same integrator, as in models with feedback or some other return action.

To combine the flexibility of the explicit integration method with the more accurate implicit methods, the predictor-corrector scheme can be adopted. The predictor, an explicit integration method, is used to predict the value of $y(k+1)$, the so-called predicted value $y^p(k+1)$. Based on this value the corresponding value of the input $f^p(k+1)$ can be calculated. Subsequently, the corrected value of the output, $y^c(k+1)$, can be made available with the aid of an implicit method.

Combination of the second order methods of Adams-Bashforth (predictor) and Adams-Moulton (corrector) yields the *Heun* integration method, namely

$$y^p(k+1) = y(k) + \frac{T}{2}\{3.f(k) - f(k-1)\} \tag{7.41a}$$

$$f^p(k+1) = f(k+1, y^p(k+1)) \tag{7.41b}$$

$$y^c(k+1) = y(k) + \frac{T}{2}\{f^p(k+1) + f(k)\} \tag{7.41c}$$

$$f(k+1) = f(k+1, y^c(k+1)) \tag{7.41d}$$

Combinations of higher order explicit and implicit integration formulas will yield a *predictor-corrector* integration method with a corresponding order.

Selection of integration interval

A major point still has to be discussed, namely the selection of an appropriate value for the integration interval T. Qualitatively, it can be stated that a small value of T will reduce the approximation errors of the numerical integration methods. On the other hand, calculation time will increase because more steps (T_f/T, with T_f the final time of a response) have to be taken. Increasing the value of T will reduce both accuracy and calculation time. As we try to achieve a sufficiently accurate result with as little calculation time as possible, a compromise has to be made.

It can also be stated that the utilization of high order Runge-Kutta methods will require more calculation time than, for example, an Adams-Bashforth method. For, a fourth order Runge-Kutta requires four calculations of $f(k)$ per integration interval in contrast with Adams-Bashforth which requires only one. A real comparison between accuracy and selection of integration interval is only possible by comparing accuracy as a function of the calculation time necessary for solving the differential equations of the model from $t = 0$ to $t = T_f$. In Figure 7.12 such a comparison has been made by using the integration methods as implemented in PSI/c for solving the second order model:

$$\ddot{y}(t) + y(t) = 0 \tag{7.42a}$$
$$\dot{y}(0) = 1 \tag{7.42b}$$
$$y(0) = 0 \tag{7.42c}$$

This model turns out to generate a sinus. Consequently, the accuracy of the calculations can be compared with the real values of $\sin(t)$, via the error criterion E

$$E = \int_0^{10} |\, y(\tau) - \sin(\tau)\,|\ d\tau \tag{7.43}$$

Calculation time T_c [s] is recorded for running this simulation model in PSI/c on a microcomputer.

Figure 7.12 illustrates the importance of selecting an appropriate value of T. Too large a value (small values of T_c) introduces approximation errors in the integration methods, and too small a value requires excessive amounts of time while the accuracy is still not attractive. This reduction of accuracy with decreasing values of T (increasing values of T_c) can be attributed to the finite accuracy of all calculations in the digital computer. The smaller the value of T, the more steps and calculations are required and the more the accuracy will suffer from roundoff errors.

Unhappily, it is rather difficult or even impossible to predict an "optimal" value for T for a given simulation model. A rule of thumb indicates that the value of T has to be about one-fifth of the smallest time constant in the model. However, the determination of the smallest time constant in a nonlinear model can be quite complicated.

A better alternative is to let an algorithm determine which values of T are

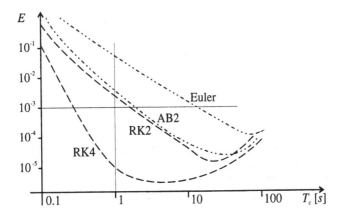

Figure 7.12 Calculation time versus accuracy

appropriate. This idea has been implemented into variable-step integration methods. Based on the value of some performance criterion E it is decided to increase, to maintain, or to reduce the step size or integration interval, as illustrated in Figure 7.13. If $E < \varepsilon_1$ the accuracy is too high (too much calculation needed) and T can be increased. If $E > \varepsilon_2$, then the accuracy suffers too much from a too large value of T. Decreasing of T has to be accomplished. The performance criterion E can be based on some measures, for instance the

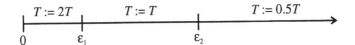

Figure 7.13 Decision about step size T

differences between $y(k + 1)$ calculated with different integration methods. A selection of the same methods but of different order is an attractive choice. A famous method is Fehlenberg's implementation of both Runge Kutta fourth and fifth order into one algorithm. The selection of some free-to-choose parameters has been done in such a way that with only six calculations of the derivative $f(k)$ in one integration interval both a fourth and a fifth order method can be executed. The difference between $y^4(k + 1)$ and $y^5(k + 1)$ gives an indication whether T can be modified. A relatively large error has to result in a reduction of T while a small error can be an indication that T should be increased.

Remark Some of these heuristic adaptive step-size methods can yield unexpected results. For, if a model reaches a steady state, all derivatives will tend to zero. Then, some methods may decide to increase the size of T, which may sometimes result in unstable responses, as illustrated in the next example. ∎

Example 7.2 Too large a selection of T may result in unstable responses of stable models. This can easily be illustrated with the aid of a first order model, solved with the aid of the Euler integration method. Based on the differential equation

$$\dot{y}(t)\tau + y(t) = K.u(t) \tag{7.44}$$

the application of Euler integration leads to the difference equation

$$y(k+1) = y(k) + T.f(k) = y(k) + \frac{T}{\tau}(K.u(k) - y(k))$$

$$= y(k).\left(1 - \frac{T}{\tau}\right) + K.\frac{T}{\tau}.u(k) \qquad (7.45)$$

For values of $T < 2\tau$ the difference equation is stable, although still large errors can be made in approximating the continuous first order model. If $T > 2\tau$ this difference equation has a pole outside the unit circle and, consequently, will exhibit divergence and instability.

For other combinations and integration methods, other boundaries are valid.

$\square$

Stiff systems

For stiff systems the integration methods presented here will yield unattractive results. In stiff systems there is a large ratio between the largest time constant τ_{max} and the smallest time constant τ_{min}, so $\tau_{max}/\tau_{min} > 100$ to 1000. Many integration steps have to be taken to achieve both accuracy ($T < \tau_{min}/5$) and a reasonable response time ($T_f > 5.\tau_{max}$), because the number of integration steps is proportional to T_f/T and so to τ_{max}/τ_{min}. For stiff systems, this ratio is large. Thus, calculations will increase if accurate answers are still expected. Stiff systems require implicit integration methods such as Gear's method, which is implemented in many simulation programs. These methods are discussed in Section 7.5.

Selection of integration method

A second observations can be made from Figure 7.12. It turns out that for this example there is a big difference between Runge-Kutta 4 and Euler. Suppose only 1 second calculation time is available. In that time Runge-Kutta 4 (RK4) yields an accuracy of about 10^{-5}, while Euler reaches about 10^{-1}, so RK4 is 10,000 times more accurate. If a certain accuracy is required, for example 10^{-3}, than RK4 needs about 0.3 seconds and Euler about 30 seconds, so RK4 can yield the same accuracy in about 1% of the time needed by Euler.

In general, a higher order integration method can yield both more accurate and faster results because it can utilize larger sizes for the integration interval T. However, an order larger than 4 to 5 turns out not to be useful. Moreover, there is a limitation in nonlinear systems. A high order polynomial is fine if the signal to be approximated, either $f(t)$ or $y(t)$, is smooth. With many nonlinearities sudden changes can be introduced which disturb high order approximations.

Although multi-step methods are very efficient (only one calculation of the derivative per integration interval), they suffer from the presence of nonlinearities and from a variable-step algorithm. For, changing the step size destroys the information of previous points of $f(k-i)$. This information has to be retrieved or recalculated.

In practice, there is a preference for using Adams-Bashforth 2, Runge-Kutta 2 and 4 and for using Runge-Kutta-Fehlenberg 4/5, which uses a variable-step mechanism. Applying Euler will almost always yield erroneous results.

7.3.3   Examples

Example 7.3 Calculate the value $y(k)$ for $k = 1$, 2 and 3 of the differential equation

$$\dot{y}(t) = t + 1 \tag{7.46}$$

with the numerical integration methods Euler, Adams-Bashforth 2 and Runge-Kutta 2 if $T = 1$ and $y(0) = 0$. For RK2 the parameters are $(a, b, p) = (\frac{1}{2}, \frac{1}{2}, 1)$.

Solution: The analytical solution $y(t) = \frac{1}{2}t^2 + t$ yields:

$$y(1) = 1.5$$
$$y(2) = 4$$
$$y(3) = 7.5$$

In this example $f(t) = t + 1$.

<u>Euler:</u>

$$y(k + 1) = y(k) + T.f(k)$$

and hence

$$
\begin{array}{rclclcl}
y(0) & = & 0 \\
y(1) & = & y(0) + T.f(0) & = & 0 + 1 & = & 1 \\
y(2) & = & y(1) + T.f(1) & = & 1 + 1.(2) & = & 3 \\
y(3) & = & y(2) + T.f(2) & = & 3 + 3 & = & 6
\end{array}
\tag{7.47}
$$

<u>Adams-Bashforth 2:</u>

$$y(k + 1) = y(k) + \frac{T}{2}(3.f(k) - f(k - 1))$$

Because Adams-Bashforth-2 is a multi-step integration method, it cannot start at $t = 0$. For in calculating $y(1)$ the value of $f(-1)$ is necessary. In general, this value is not defined and so not available. Consequently, Euler is selected for calculating $y(1)$.

$$
\begin{array}{rclcl}
y(1) & = & y(0) + T.f(0) & = 1 \\
y(2) & = & y(1) + \frac{T}{2}\{3.f(1) - f(0)\} & = 1 + \frac{\{3.2-1\}}{2} & = 3.5 \\
y(3) & = & y(2) + \frac{T}{2}\{3.f(2) - f(1)\} & = 3.5 + \frac{\{3.3-2\}}{2} & = 7
\end{array}
\tag{7.48}
$$

This example illustrates that Euler introduces relatively large errors. Although Adams-Bashforth-2 has second order accuracy and $y(t)$ is second order, Adams Bashforth-2 still yields erroneous results, because $y(1)$ has been calculated by the first order Euler integration method. In using Runge Kutta-2, a second order one-step integration method, accurate results can be expected.

<u>Runge-Kutta 2:</u>

With $(a, b, p) = (\frac{1}{2}, \frac{1}{2}, 1)$ RK2 becomes:

$$y^p(k + 1) = y(k) + T.f(k) \tag{7.49a}$$
$$f^p(k + 1) = (k + 1) + 1 \tag{7.49b}$$
$$y(k + 1) = y(k) + \frac{1}{2}T\{f(k) + f^p(k + 1)\} \tag{7.49c}$$
$$f(k + 1) = (k + 1) + 1 \tag{7.49d}$$

Hence:

$$
\begin{array}{rcllcl}
y^p(1) & = & y(0) + f(0) & = & 0 + 1 & = & 1 \\
f^p(1) & = & 1 + 1 & = & 2 & & \\
y(1) & = & y(0) + \frac{T}{2}\{f(0) + f^p(1)\} & = & 0 + \frac{1}{2}\{1 + 2\} & = & 1.5 \\
y^p(2) & = & y(1) + \hat{f}(1) & = & 1.5 + 2 & = & 3.5 \\
f^p(2) & = & 2 + 1 & = & 3 & & \\
y(2) & = & y(1) + \frac{T}{2}\{f(1) + f^p(2)\} & = & 1.5 + \frac{1}{2}\{2 + 3\} & = & 4 \\
y^p(3) & = & y(2) + \hat{f}(2) & = & 4 + 3 & = & 7 \\
f^p(3) & = & 3 + 1 & = & 4 & & \\
y(3) & = & y(2) + \frac{T}{2}\{f(2) + f^p(3)\} & = & 4 + \frac{1}{2}\{3 + 4\} & = & 7.5
\end{array}
$$

□

Example 7.4 Calculate the value of $y(k)$, $k = 1, 2$, of the second order system of Figure 7.14 with $u(t) = 1$ for $t \geq 0$, $w(0) = 0$ and $y(0) = 0$ with the numerical integration method Euler with $T{=}1$.

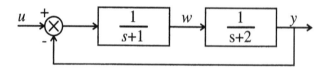

Figure 7.14 Second order system

Clearly, this step size is too large for obtaining accurate results. Still, this value is selected to reduce the computational complexity in this example.

Solution: The following equations can be derived:

$$
\dot{y} = w - 2y\dot{w} = u - y - w \tag{7.50}
$$

Consequently, the derivative functions f_y and f_w become:

$$
f_y(k) = w(k) - 2y(k)f_w(k) = u(k) - y(k) - w(k) \tag{7.51}
$$

Euler:

$$
\begin{array}{rcllcl}
y(1) & = & y(0) + T.f_y(0) & = & 0 & & \\
w(1) & = & w(0) + T.f_w(0) & = & w(0) + u(0) - y(0) - w(0) & = & 1 \\
y(2) & = & y(1) + T.f_y(1) & = & y(1) + w(1) - 2y(1) & = & 1 \\
w(2) & = & w(1) + T.f_w(1) & = & w(1) + u(1) - y(1) - w(1) & = & 1
\end{array}
$$

□

7.4 *Conversion from parallel to series*

A continuous or discrete simulation model is assumed to be parallel. All variables change their values simultaneously. Solving a set of equations on a digital computer, which has only one or a finite number of calculating elements, requires some kind of ordering. There has to be a calculating sequence. This

sequence has to preserve the parallel character of the model although the calculations are calculated sequentially.

First, the influence of the sorting procedure will be shown. Suppose that
the simulation model consists of three gain elements, A, B and C, having gains
1, 2 and 3 respectively. This model is illustrated in Figure 7.15.

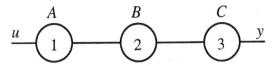

Figure 7.15 Three gains

At $t = 0$, an input variable u becomes active; the input becomes $u = 1$. For
$t = 0$, all gains have an initial value 0. Two different calculating sequences will
be discussed, namely u, A, B, C and u, C, B, A. The results of the calculations,
depending on the sorting sequence are illustrated in Table 7.3.

Table 7.3
Influence of the sorting sequence

u	A	B	C	Calculating sequence	u	C	B	A
A	=	u			C	=	3	*	B
B	=	2	*	A	B	=	2	*	A
C	=	3	*	B	A	=	u		

u	A	B	C	Period	u	A	B	C
1	1	2	6	0	1	1	0	0
1	1	2	6	1	1	1	2	0
1	1	2	6	2	1	1	2	6

In both cases A receives the value 1 in period 0. The value of B becomes
$2.A$. In the first situation this becomes 2, but in the second situation B remains 0
because A has not yet received its value 1 from u. Using the second calculating
order introduces a dead time of $2T$ in the model. This dead time does not exist
in the original model, so it is the result of solving this problem on a sequentially
oriented computer.

This example is simple and straightforward. The appropriate sorting sequence
can easily be determined. In cases where feedback is present, the determination
of an appropriate calculating sequence is not simple. In simulation programs
and languages sorting procedures have been implemented to take care of this
sorting sequence. The basics of each sorting sequence are to add only blocks to
a list of blocks if they satisfy the following conditions:

- a block can only be calculated if all its inputs are known;

- an input is known if the associated block is either

    - already included in the list of accepted blocks or

– it is a block with no immediate relation between input and output, for example an integrator, delay, dead time, transfer function with more poles than zeros, etc.

The assumption concerning integrators is that the integration method is an explicit one. Only then it is possible to calculate the first order model of Figure 7.16. With explicit integration methods the value of the output at t=k+1 does not de-

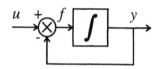

Figure 7.16 First order model

pend on the value of the input at t=k+1. So, there is no immediate reaction of $y(k + 1)$ to variations of $f(k + 1)$, which allows integrators to have their own output value as input.

Sometimes it is not possible to determine an appropriate sorting sequence. Then an algebraic loop has been detected. In an algebraic loop each block or function represents some linear or nonlinear algebraic function. These functions have an immediate relation between their input and output. An example of an algebraic loop is illustrated in Figure 7.17.

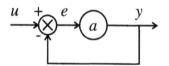

Figure 7.17 Algebraic loop

Suppose that this loop is calculated via the following equations:

$$e = u - y \tag{7.52}$$

$$y = a.e$$

At $t = 0$, we assume that $e = y = 0$ and that $u = 1$. Then the following values can be calculated:

$$
\begin{array}{llll}
e & = 1 - 0 & = 1 \\
y & = a.1 & = a \\
e & = 1 - a \\
y & = a.(1 - a) & = a - a^2 \\
e & = 1 - (a + a^2) \\
y & = a.e & = a - a^2 + a^3
\end{array}
\tag{7.53}
$$

Or, for large values of the time:

$$y = a - a^2 + a^3 - a^4 + a^5 - \ldots \tag{7.54}$$

Only for -1 < a < 1, will this sequence converge to the required value

$$y = \frac{a}{1 + a} \tag{7.55}$$

Values of a outside this region will make this loop diverge or explode.

Another possible way of analyzing this model is to describe it with the aid of difference equations. It will become clear that the equations have to be interpreted as:

$$e(k) = u(k) - y(k-1) \qquad (7.56)$$
$$y(k) = a.e(k)$$

The value of $e(k)$ depends on the old value of y, because the new one has not yet been determined. Using the Z-transformation yields

$$\frac{Y(z)}{U(z)} = \frac{a.z}{z+a} \qquad (7.57)$$

Instead of having an algebraic function between u and y, the fact that the equations have to be solved sequentially introduces dynamics into an algebraic loop. The responses of $y(k)$ to a unit step of $u(k)$ are given in Figure 7.18 for several values of a. Clearly, an algebraic loop can, if not detected, introduce instability in a stable model.

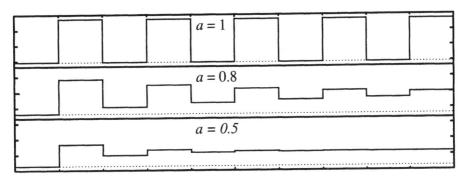

Figure 7.18 Oscillations due to an algebraic loop

There are several ways of solving algebraic loops. The best method is to avoid them in a model description as much as possible. This approach is not always possible. Then, if the loop gain is small (smaller than 1), a unit delay can be introduced in the model. Although this extra delay changes the model behavior, these changes may be acceptable. The last method for solving algebraic loops is to solve them iteratively with appropriate methods. Figure 7.19 illustrates the approach of the PSI/c method.

In the algebraic loop an additional block ALB is introduced. Its input is called $g(x(k))$ and its output $x(k)$. The purpose of this block is to search at $t = t_k$ for a value of $x(k)$ that will yield a value $g(x(k))$ that is equal to $x(k)$. In that case the loop has been solved and the effect of the block on the simulation model can be neglected. The value of $g(x(k))$ depends completely on the other blocks that shape the algebraic loop.

Searching at $t = t_k$ for a value of $x(k)$ that makes $g(x(k)) = x(k)$ is effectuated via the Newton-Raphson iterative method of looking for a root $x(k)$ of the equation $G(x(k)) = 0$:

$$G(x(k)) = g(x(k)) - x(k) = 0 \qquad (7.58)$$

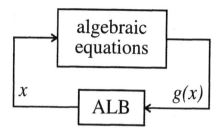

Figure 7.19 Algebraic loop solution in PSI

via

$$x^{i+1}(k) = x^i(k) - \dot{G}^{-1}(x^i(k)).G(x^i(k)) \qquad (7.59)$$

with

$$\dot{G}(x^i(k)) = \frac{G(x^i(k) + \Delta x) - G(x^i(k))}{\Delta x} \qquad (7.60)$$

with Δx a small value. The Newton-Raphson method is very fast in determining a root of a nonlinear equation. However, sometimes the method does not converge. For these situations, if no convergence has been detected, a slower, but more reliable method is used, namely by solving the optimization problem:

$$x(k) = \arg \min_{x(k)} |g(x(k)) - x(k)| \qquad (7.61)$$

The index i of $x^i(k)$ indicates the iteration number. At any point in time t_k determined by the numerical integration method, the search algorithm for determining a solution of $G(x^i(k)) = 0$ has to be started. If the integration interval T is small and the variables do not change rapidly, the solution $x(k-1)$ of a previous integration interval is a good starting point $x^0(k)$ for the present integration interval. Then, the solution can be obtained quickly. Otherwise large amounts of calculation time have to be spent in searching for an appropriate value of $x(k)$.

7.5 *Solving stiff ODEs and DAEs*

The numerical integration methods discussed up to now are intended for solving ordinary differential equations, ODEs. These ODEs are formulated as explicit differential equations such as

$$\dot{y}(t) = f(y(t), t) \qquad (7.62)$$

These integration methods are called explicit integration methods because they require explicit knowledge of $\dot{y}(t)$. Their integration interval has to be chosen small $T < \tau_{min}/5$ to preserve stability and accuracy. Consequently, they are not suited for solving stiff differential equations (too much calculation time required) or DAEs (because in $F(\dot{y}(t), y(t), t) = 0$, $\dot{y}(t)$ is not explicitly known).

In the following example, the difference between explicit and implicit methods is explained.

Example 7.5 (Explicit and implicit methods)

Explicit methods

Explicit numerical integration methods lack stability if T is selected too large. Suppose $\dot{y}(t) = -ay(t)$, with a a positive constant. This model is stable. The explicit Euler method with integration interval T yields:

$$y(k+1) = y(k) + T\dot{y}(k) = y(k) - Tay(k) = (1 - aT)y(k) \qquad (7.63)$$

This difference equation is only stable if $T < 2/a$.

Implicit methods

If the implicit Euler method is utilized for solving the same model $\dot{y}(t) = -ay(t)$, the following difference equation arises

$$y(k+1) = y(k) + T\dot{y}(k+1) = y(k) - Tay(k+1) \qquad (7.64)$$

This difference equation is always stable for any $T > 0$. So, even for large values of T, stability, although not accuracy, is guaranteed. ⬛

All implicit numerical integration methods yield better stability properties. However, $\dot{y}(k+1)$ is not available, so it has to be approximated. Suppose $\dot{y}(t) = f(y(t), t)$. Then, the implicit Euler algorithm yields

$$y(k+1) = y(k) + T.\dot{y}(k+1) = y(k) + T.f(y(k+1), k+1) \qquad (7.65)$$

Linearization   Linearization of the nonlinear equation (7.65) yields

$$
\begin{aligned}
y(k+1) &= y(k) + T.f(y(k+1), k+1) \\
&= y(k) + T[f(y(k), k+1) + \frac{\partial f}{\partial y}.(y(k+1) - y(k)] \\
&= y(k) + T[f(y(k), k+1) + J.(y(k+1) - y(k)]
\end{aligned}
$$

with J the Jacobian of $f(y)$, the matrix of partial derivatives of f with respect to $y(k)$. Then, $y(k+1)$ becomes

$$y(k+1) = y(k) + T(I - T.J)^{-1}.f(y(k), k+1) \qquad (7.66)$$

So, the value $y(k+1)$ is calculated with knowledge of both f and J, evaluated at $(y(k), k+1)$. For nonlinear models the Jacobian J depends on $y(k)$, so at any point in time k the Jacobian J needs to be evaluated. Calculation time is saved if the recalculation of J, and the calculation of $(I - T.J)^{-1}$ are executed only a limited number of times.

In this section the low order implicit Euler method is used. In the next section, methods are discussed to calculate $y(k+1)$ with higher order implicit integration methods. These higher order methods are necessary to solve DAEs accurately. By an error analysis it is shown that still errors can be expected if higher index DAEs are solved.

7.5.1 BDF methods

The set of nonlinear equations (7.65) can also be solved iteratively, for example by a Newton iteration: estimate $y^i(k+1)$, calculate $f(y^i(k+1), k+1)$, etc. The

Newton method is one of the fastest methods for solving a set of nonlinear equations.

So, $y^i(k + 1)$ is calculated based on all information up to and including $t = k.T$. The BDF-method (Backwards Difference Formulas) replaces $\dot{y}(k + 1)$ by, for example, the first order approximation $(\hat{y}^i(k + 1) - y(k))/T$. Then the original set of differential equations

$$F(\dot{y}(k + 1), y(k + 1), k + 1) = 0 \tag{7.67}$$

is replaced by

$$F'(y^i(k + 1), y(k), k + 1) = 0 \tag{7.68}$$

with $y(k)$ the known and $y^i(k + 1)$ the unknown variables.

As a consequence, the differential equation is replaced by a set of nonlinear algebraic equations with unknowns $y^i(k + 1)$. This set of equations is solved iteratively by the Newton method.

$$y^{i+1}(k + 1) = y^i(k + 1) - J^{-1}F'(y^i(k + 1), y(k), k + 1) \tag{7.69}$$

This procedure consists of two steps.

- First, calculate an estimate for $y^0(k+1)$ with e.g. the implicit Euler method or another BDF method.

- Second, apply a Newton-Raphson iteration to solve the resulting set of nonlinear algebraic equations. If $F'(y^i(k + 1), y(k), k + 1) < \epsilon$, with ϵ a small positive constant, then a solution for $k + 1$ has been found.

The same approach is used in solving DAEs. Solving the nonlinear algebraic equations with the Newton-Raphson method requires knowledge of the derivative (Jacobian) $J = dF/dy$, of F with respect to $y(k + 1)$. Therefore, the solution of DAEs or stiff differential equations is a time-consuming activity.

7.5.2 Error analysis of a BDF method

In this section a higher order implicit integration method is discussed. This method uses more values of $y(k)$ in the past. These values $y(k)$ need not to be located at equal distances in time. So, the integration interval T_k becomes dependent of k. The numerical solution of the DAE

$$F(\dot{y}, y(t), t) = 0 \tag{7.70}$$

by a BDF method is based on the approximation of $y(t)$ and $\dot{y}(t)$ by an m-th order polynomial $p(t)$. This polynomial passes through $m + 1$ old values of $y(t_i), i = 0, m$. So, $p(t) = y(t)$ at these $m + 1$ points. The *Newton interpolation formula* yields

$$p(t) = c_0 + c_1(t - t_0) + c_2(t - t_0)(t - t_1) + \ldots + c_m(t - t_0)(t - t_1) \ldots (t - t_{m-1}) \tag{7.71}$$

The coefficients $(c_0, \ldots, c_m)$ are calculated recursively by a simple substitution from the following system of linear equations:

$$
\begin{aligned}
y(t_0) &= c_0 \\
y(t_1) &= c_0 + c_1(t_1 - t_0) \\
y(t_2) &= c_0 + c_1(t_2 - t_0) + c_1(t_2 - t_0)(t_2 - t_1) \\
&\vdots \\
y(t_m) &= c_0 + c_1(t_m - t_0) + \ldots + c_m(t_m - t_0)(t_m - t_1) \ldots (t_m - t_{m-1})
\end{aligned}
$$

These coefficients c_i determine $p(t)$ uniquely and are calculated with the values of $y(t_i), i = k, k - m$ and the values of $t_i, i = k, k - m$. It can be shown (Gear, 1974) that $\dot{p}(t)$ satisfies

$$\dot{p}(k) = -\frac{1}{T_k} \sum_{i=0}^{m} \alpha_i y(k-i) = \frac{1}{T_k} \rho_k y(k) \tag{7.72}$$

with $\rho_k y(k)$ a short-hand notation of the finite sum. The values of α_i are calculated with the values of c_i.

If $y(t)$ has continuous derivatives of order at least $m+1$ then the interpolation error $e(t)$ between $y(t)$ and the interpolation $p(t)$ becomes (Gear, 1974)

$$e(t) = p(t) - y(t) = \frac{y^{(m+1)}(\xi)}{(m+1)!}(t - t_0)(t - t_1)\ldots(t - t_m) \tag{7.73}$$

where $\xi \in [\min(t, t_0), \max(t, t_m)]$.

The error $\dot{e}(t)$ in the estimation $\dot{p}(t)$ of the derivative $\dot{y}(t)$ is calculated as:

$$\dot{e}(t) = \frac{y^{(m+1)}(\xi)}{(m+1)!} \cdot \frac{d}{dt}((t - t_0)(t - t_1)\ldots(t - t_m)) \tag{7.74}$$

which, when evaluated at one of the mesh points $t_0, \cdots, t_m$, for example at point t_0, equals

$$\dot{e}(t_0) = \frac{y^{(m+1)}(\xi)}{(m+1)!} \cdot (t_0 - t_1)\ldots(t_0 - t_m) \tag{7.75}$$

The *truncation error* τ of the BDF method is defined as

$$\tau = -T_k \dot{e}(k) \tag{7.76}$$

Hence

$$\tau = -T_k \frac{y^{(m+1)}(t_k)}{(m+1)!} \prod_{j=1}^{m}(t_k - t_{k-j}) + O(T^{m+2}) \tag{7.77}$$

We can conclude from (7.77) that:

- For a fixed step-size: $\tau = -\frac{1}{m+1}T^{m+1}y^{(m+1)}(t_k) \sim O(T^{m+1})$

- For a varying step-size $\tau \sim O(T_{max}^{m+1})$

- For a varying step-size $\tau \sim O(T_k^2)$ if old steps $T_{k-1}\ldots T_{k-m}$ are fixed

Remark In many codes of computer programs, algorithms with varying step-size T_k are implemented. Based on an estimation of the local truncation error τ the step-size T_k is adjusted, while the previous ones $T_{k-1}\ldots T_{k-m}$ are constant or fixed. Then, the truncation error satisfies $\tau \sim O(T_k^2)$ ∎

7.5.3 *Propagation of the truncation error*

In this section we estimate the error $e_k = y_k - y(t_k)$ in the numerical solution of a fully implicit DAE:

$$F(\dot{y}, y, t) = 0 \tag{7.78}$$

The numerical solution y_k of (7.78) is calculated by using the BDF algorithm

$$F\left(\frac{\rho_k y_k}{T_k}, y_k, t\right) = 0 \qquad (7.79)$$

Suppose that $y(t_k)$ represents the true solution at t_k. Then

$$\frac{\rho y(t_k)}{T} - \dot{y}(t_k) = \frac{\tau_k}{T} \qquad (7.80)$$

where τ_k is the truncation error at t_k. Then

$$0 = F\left(\frac{\rho y_k}{T}, y_k, t_k\right) \qquad (7.81a)$$

$$= F\left(\frac{\rho y(t_k)}{T} + \frac{\rho e_k}{T}, y(t_k) + e_k, t_k\right) \qquad (7.81b)$$

$$= F\left(\dot{y}(t_k) + \frac{\tau_k}{T} + \frac{\rho e_k}{T}, y(t_k) + e_k, t_k\right) \qquad (7.81c)$$

A Taylor expansion around point $(\dot{y}(t_k), y(t_k), t_k)$ yields

$$0 = F(\dot{y}(t_k), y(t_k), t_k) + \frac{\partial F}{\partial y} e_k + \frac{\partial F}{\partial \dot{y}}\left(\frac{\tau_k}{T}\right) + \frac{\partial F}{\partial \dot{y}}\left(\frac{\rho e_k}{T}\right) \qquad (7.82)$$

with notation: $E_k = \partial F / \partial \dot{y}$, $A_k = \partial F / \partial y$ we obtain

$$E_k \rho e_k + T A_k e_k + E_k \tau_k = 0 \qquad (7.83)$$

Solving this for e_k we get

$$e_k = (E_k + T A_k)^{-1} E_k \sum_{i=1}^{m} \alpha_i e_{k-i} + (E_k + T A_k)^{-1} E_k \tau_k \qquad (7.84)$$

Equation (7.84) indicates that the matrix term $(E_k + T A_k)^{-1}$ rules the behavior of numerical errors.

If we combine the estimation of e_k (7.84) with the estimation of the truncation error τ_k (7.77), we find for algorithms with a variable step size

$$e_k \sim (E_k + T A_k)^{-1} E_k \sum_{i=1}^{m} \alpha_i e_{k-i} + (E_k + T A_k)^{-1} E_k \cdot O(T^2) \qquad (7.85)$$

Even with $e_{k-i} = 0$, e_k can be nonzero if $(E_k + T A_k)^{-1} E_k \cdot O(T^2)$ cannot be neglected for small values of T. It would be very convenient if the appearance of T in the matrix $(E_k + T A_k)^{-1} E_k$ were known. Then, an estimate of the error e_k would become available and explains whether the value of e_k will decrease for decreasing values of the step size T. Graph-oriented methods are introduced to determine the power of s or T of each entry of the matrix $(E_k + T A_k)^{-1} E_k$.

For index 0 and 1 DAE, the error e_k will vanish, for index 3, the error does not decrease for decreasing values of T, but remains constant.

7.5.4 Newton iteration

The Newton method is used in solving sets of nonlinear algebraic equations $F = 0$.

$$y^{i+1}(t_k) = y^i(t_k) - J^{-1}F(y^i(t_k), t_k, T) \tag{7.86}$$

with J the Jacobian of F, namely

$$J = \frac{dF}{dy} = \frac{\partial F}{\partial y} + \frac{\partial F}{\partial \dot{y}}\frac{\partial \dot{y}}{\partial y} = \frac{\partial F}{\partial y} + \frac{1}{T}\frac{\partial F}{\partial \dot{y}} \tag{7.87}$$

If the linear constant DAE is defined by

$$E\dot{y}(t) + Ay(t) = f(t) \tag{7.88}$$

then the Jacobian J becomes $J = A + E/T = (E + TA)/T$ and, consequently, the inverse Jacobian $J^{-1} = T(E + TA)^{-1}$.

With an index 1 DAE , some of the entries in the inverse Jacobian J^{-1} are constant and others are proportional to positive powers of T, so with decreasing values of T, the inverse Jacobian in (7.86) yields no problems. If the index is 3, some of the entries in the inverse Jacobian J^{-1} are constant and others are proportional to T^{-2}. With decreasing values of T, the inverse Jacobian in (7.86) becomes very large, which makes it impossible to locate an accurate solution of the nonlinear algebraic equation.

7.5.5 Numerical problems of higher index DAE

In several sections the numerical behavior of the solution of DAEs has been analyzed. Combining these sections the following remarks can be made.

- In Section 7.5.3 an estimation was made of the local error, especially the dependency of this error on decreasing values of the step size T. For index 3 problems, the error does not vanish for decreasing values of T. Consequently, a step-size algorithm expects a decreasing error for decreasing values of T. If this is not recognized, very small values of T may arise. This will conflict with another observation.

- In Section 7.5.4, we showed that the inverse Jacobian J will become very large for small values of T. Then, the convergence is no longer guaranteed, which yields inaccurate or even "explosive" solutions.

- At $t = 0$ a serious initialization problem is present in DAEs. Each independent variable will have its own initialization. However, in a DAE, the independent variables are coupled by algebraic equations. It requires much calculation to guarantee that the solution is a consistent set.

These observations indicate that index 3 problems cannot be solved reliably. Even the solution of index 2 DAE problems require advanced numerical methods and fine tuning. The program *DASSL* is guaranteed to give accurate and reliable results for index 0 and 1 DAE problems. In general, index 2 problems are solved, but index 3 problems are either refused, the solution explodes or inaccurate results are achieved.

There are several remedies that can be used.

- The modeler can take care of his/her model. Reduce possible higher order situations as much as possible; for example combining two rigid connected masses, two inductors in series, two capacitors in parallel.

- By applying symbolic manipulation, remove algebraic constraints as much as possible. Replace them by their differentiated equation.

- A disadvantage of using differentiated constraints is that the original constraint is violated owing to numerical errors. If there is no mechanism to control this drift, meaningless results are obtained. To counteract the drift, additional equations and variables (Lagrange multipliers) can be added to take care of the original constraint.

- By applying a weighing in the error calculation of the varying step-size algorithm, excessively small values of T with unavoidable numerical errors in solving the nonlinear algebraic equation, are avoided.

7.6 *Applications*

Simulation is a general purpose tool for solving technical and nontechnical problems. In previous sections, examples of a prey-predator model, an economic model, heating a house, etc, have been given. They illustrate the broad applicability of simulation techniques.

Two dedicated applications of simulation combined with optimization concern parameter estimation and controller design of nonlinear models.

7.6.1 *Parameter estimation*

Suppose that measurements of input $u(k)$ and output $y(k)$ are available. Then the question is posed which model describes as accurately as possible the relation between input and output as given by the measurements $\{u(k), y(k)\}_N$. The expression "as accurately as possible" already indicates that a search has to be made for the best model. A good model has an output $y_m(k)$, calculated on the basis of the same input sequence $u(k)$, that can be compared with the measured output $y(k)$. A measure can be, for example, the sum of the absolute values between $y(k)$ and $y_m(k)$ for all measurements k, or

$$J(\theta) = \sum_{k=1}^{N} \mid y_m(k, \theta) - y(k) \mid \qquad (7.89)$$

The best or optimal estimate $\hat{\theta}$ of the parameters θ is given as:

$$\hat{\theta} = \arg \min_{\theta} J(\theta) \qquad (7.90)$$

It is the user's responsibility to select an appropriate model: order, structure and unknown parameters θ. Because $y_m(k, \theta)$ is calculated by simulation, any model can be utilized, such as continuous, discrete, nonlinear, etc. The minimization problem can be solved by nonlinear optimization methods (Chapter 8). The proposed approach is illustrated in Figure 7.20.

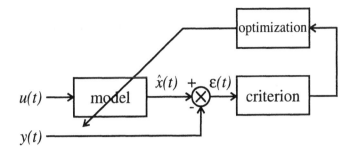

Figure 7.20 Parameter estimation via simulation and optimization

The power and flexibility of this approach to parameter estimation lie in the possibility of using all kinds of criteria to define a fit between model and measurements, and the capability of simulation to utilize almost any model description. Moreover, a priori information concerning parameter values or model structure can be implemented and fully exploited.

7.6.2 *Controller design*

A second important application of both simulation and optimization is in the design of controllers. It can be stated that a requirement for any design will be that the output $y(t)$ of a controlled system follows the required reference value $r(t)$ as well as possible. A criterion to measure this ability to follow $r(t)$ can be found by defining one that is based on the error signal $e(t) = y(t) - r(t)$. For example, $J(\theta)$ is an appropriate criterion with θ the parameters of a controller:

$$J(\theta) = \int_0^{T_f} \{y(\tau, \theta) - r(\tau)\}^2 \, d\tau \qquad (7.91)$$

By applying simulation it is possible to calculate the value of this criterion for some selected set of parameters θ of the controller. A nonlinear optimization technique is able to find values θ^* of the controller parameters that yield the lowest value of the criterion, so

$$\theta^* = \arg \min_\theta J(\theta) \qquad (7.92)$$

If this iterative optimization process is completed, "optimal" values of the controller parameters have been found. This setup is illustrated in Figure 7.21.

In Figure 7.22 the step responses of a third order model with dead time are shown, if this model is controlled by means of an "optimal" P- and PI-controller.

It has to be emphasized that the proposed optimization procedure does not determine a controller structure. The designer has to select such a structure, after which the optimization will try to find the "optimal" values.

Remark Note that besides the selection of the controller structure (order, feedback and/or feedforward, etc.) also other choices determine the "optimal" solution. For example, the choice of the criterion or cost function (Integral square error, integral absolute error, etc. the length (T_f) of the simulation run), the selection of the shape and size of $r(t)$, the presence of disturbances, etc. ∎

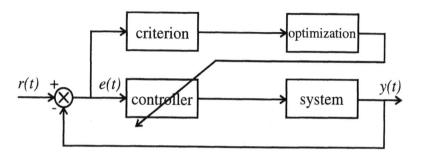

Figure 7.21 Controller design by simulation and optimization

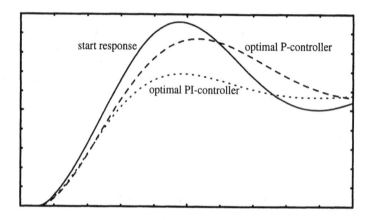

Figure 7.22 Responses of optimal controllers

7.7 *Summary*

In this chapter we discuss techniques for using simulation as a tool for solving sets of difference and/or differential equations. Especially, the digital computer is well suited for executing a simulation. It is shown that there are three important topics that influence the accuracy of a simulation. First there is the accuracy of the calculations on a digital computer. With floating-point numbers accuracy can be achieved. Second, a model is formulated as a set of parallel equations that are solved with the aid of a serial digital computer. Sorting of all calculations is required to avoid errors. If algebraic loops are present, appropriate sorting is not possible. The solution of an algebraic loop is obtained by using an algorithm for solving sets of nonlinear algebraic equations. The third topic that has to be considered is the solution of differential equations with the aid of a digital computer. Only the application of numerical integration methods provide sufficient flexibility to solve any set of nonlinear differential equations. Numerical integration methods require an explicit formulation of first order differential equations, such as $\dot{y}(t) = f(y(t), t)$.

The time axis is divided into parts of size T, the integration interval. By taking the values of f at several points in time, and the values of y(t) in the past, numerical integration methods find an approximation of y(t) at future points in time. The accuracy of this approximation is determined by the selection of the integration method and the size of the integration interval T. With a method of high order and T small, accurate results can be expected. Variable step size numerical integration methods adjust the size of T according to some error criterion.

Explicit methods use only information from the past. Their stability for large step sizes is poor. Implicit methods perform better, but can be more time-consuming. For the solution of stiff differential equations and for differential algebraic equations, they are to be preferred.

7.8 References

Simulation as a general methodology is discussed in, e.g., Kheir (1988).

The numerical aspects of solving ordinary differential equations are treated in many books on numerical analysis.

We only mention the basic contributions of Gear (1971), Brenan, Campbell and Petzold (1989) and Hairer, Lubich and Roch (1989). These books especially focus on the solution of stiff ordinary differential equations and differential algebraic equations.

7.9 Problems

1. Calculate the first four values of the output $y(k)$ of a first order model with gain 1 and time constant 1 with the Euler integration method if the integration interval is selected as $T = \frac{1}{2}$.

2. Derive a simulation diagram with only first order integral and algebraic equations of the linear model $H(s) = \dfrac{(s+2)}{(s^2 + 3s + 1)}$.

3. Describe three different limitations of the digital computer if it is used for the simulation of continuous models. Does an analog computer have the same limitations?

4. Calculate the values of $y(t)$ for $t = 1$ and 2 seconds with the aid of the Adams-Bashforth 2 integration method. The output $y(t)$ arises when a unit step is applied to $H(s)$:

$$H(s) = \frac{2}{10s + 1} \tag{7.93}$$

Assume an integration interval T of 1 second.

5. Explain Figure 7.12 by answering the following questions:

   - why is the accuracy of RK4 better than Euler if the same amount of calculation time is used?
   - why does the accuracy of RK4 improve for increasing values of T_c until $T_c = 10s$, and why does the accuracy reduce for $T_c > 10s$?

6. Determine an appropriate calculation sequence of the block diagrams in Figure 7.23.

   Determine whether an algebraic loop will be present in these models if the integration method is explicit or implicit.

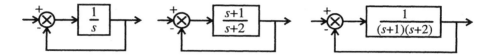

Figure 7.23 Simulation models

7. The numerical integration methods developed in Chapter 7 are based on equal step sizes T. With multi-step methods this imposes limitations if a variable step algorithm is utilized or if state- or discrete-events occur. Consequently, it is useful to derive a multi-step numerical integration method based on different values $T(k)$ for each interval k.

   (a) Derive the numerical integration methods of Euler and Adams-Bashforth 2 with $T(k)$ not constant.

   (b) Calculate $y(k + 1)$ with these formulas if $T(k - 1) = 1$, $T(k) = 2$, $y(k) = 1$, $f(k - 1) = 2$ and $f(k) = 3$.

8. A model consists of a gain a with unity feedback, so $y = a.e$ and $e = r - y$. The model is calculated with Euler as numerical integration method with integration interval T=1. The simulation program detects an algebraic loop. Still, the program tries to solve this loop by assuming a delay in the feedback loop ($e(k) = r(k) - y(k - 1)$). For $t < 0$ the model is in rest ($r = y = 0$). For $t \geq 0$ r has the value 1, so $r(t) = 1, t \geq 0$).

   (a) Calculate the values of $y(t)$ for $a = \frac{1}{2}$, for $a = 1$ and for $a = 2$ if t=0, 1, 2 and 3 sec.

   (b) Does the solution depend on the size of the integration interval T?

   (c) Does the solution depend on the selected integration method?

   (d) The solution found is not exact. How can you solve the algebraic loop accurately? Does it matter if $|a|$ is smaller than 1? Explain your answers.

chapter eight

Optimization

8.1 *Introduction*

In the preceding chapters, optimization is introduced for solving parameter estimation and controller design problems. However, optimization is a widely used approach for solving not only technical, but also economical and financial problems. A problem is posed as a verbal expression, for example:

- Search for the parameters of a model such that the output $y_m(k)$ of that model for some input signal $u(k)$ coincides "as accurately as possible" with the measured output signal $y(k)$ of a system.

- Determine the parameters of a controller such that the closed-loop system behaves "as accurately as possible".

- Design a controller for a room thermostat that yields the "lowest" energy consumption while the temperature in the room is at least 19 $^\circ$C.

- Determine the "best" size of a ship to carry goods between harbor A and harbor B.

- Which car is "best" suited for my needs?

- Determine the profit rate such that the national income increases "as fast as possible".

The first problem resembles the parameter estimation problem as introduced in Chapters 3 to 5. The following two examples deal with the design of an "optimal" controller, while the other examples are taken from nontechnical problems. To realize a tractable mathematical problem, verbal expressions such as "as accurately as possible" and "lowest" have to be formulated in a mathematical framework as a mathematical norm. For each selection of the design parameters a single function value has to express "how satisfactory" this selection has been. The selection that yields the lowest value of this function is called the optimal solution.

The parameter estimation problem can be converted into a mathematical optimization problem. Searching for the "best" parameters is translated into finding parameter values which yield the lowest possible value of a cost function. Many different optimization techniques exist for solving almost any mathematical optimization problem. Therefore, optimization is a powerful and flexible tool in many applications. In this chapter, it will be treated in greater detail and both its power and its limitations will be further explained.

Any properly defined optimization problem consists of three separate characteristics, namely:

- $J(\theta)$: a cost function or criterion that expresses the intention or goal;

- θ: the parameters that can be used to optimize the criterion $J(\theta)$;

- $h(\theta)$: the constraints in the parameter space ($h(\theta) \leq 0$) that bounds the solution to the feasible or allowed region.

The first two elements, namely $J(\theta)$ and θ, are always present in an optimization problem. Assuming that we are searching for J minimum, the optimal solution θ^* is formulated as:

$$\theta^* = \arg\min_{\theta} J(\theta) \tag{8.1}$$

So,

$$J(\theta^*) < J(\theta) \qquad \forall \theta, \ \theta \neq \theta^* \tag{8.2}$$

The third element, $h(\theta)$, can be omitted if no constraints are imposed on the values of the parameters. It has to be noted that a modification of the criterion and/or of the parameters and/or of the constraints will yield another optimization problem and, consequently, another solution that is also optimal.

Without loss of generality it is assumed that all optimization problems are minimization problems, for

$$\max_{\theta} J(\theta) = \min_{\theta} -J(\theta) \tag{8.3}$$

The selection of the criterion is, in general, a compromise between the desire to have an appropriate measure to express the desired goal (fitting measurements, good control behavior) and the potential of optimization techniques to locate the optimal solution. For example, the Least Squares method utilizes the sum of squares of the prediction error $\epsilon(\theta, k)$, so that

$$J(\theta) = \frac{1}{N} \sum_{k=1}^{N} \epsilon^2(\theta, k) \tag{8.4}$$

Clearly, the cost function or criterion $J(\theta)$ of the parameter vector θ, with $\theta = \begin{pmatrix} \vartheta_1 & \cdots & \vartheta_n \end{pmatrix}^T$, is selected as the sum of the squares of the prediction error $\epsilon(\theta, k)$. This specific selection of $J(\theta)$ is made to ensure an easy and fast solution of the optimization problem. Another selection of $J(\theta)$, with

$$J(\theta) = \sum_{k=1}^{N} |\epsilon(\theta, k)| \tag{8.5}$$

yields a different optimization problem and correspondingly a different solution and is, sometimes, more difficult to solve.

Besides having an easy solution to the optimization problem, it is necessary to ensure that the located minimum really reflects the optimal solution of the problem. This requirement means that the optimization problem must be unimodal, i.e. that the problem has only one optimum. Then any optimum is a global optimum. Local optima do not exist, so any minimum located represents the solution of the optimization problem. In Figure 8.1 a convex, a unimodal and a nonunimodal function $J(\theta)$ are illustrated. A function $J(\theta)$ is convex if it satisfies

$$J(\lambda\theta_1 + (1 - \lambda)\theta_2) < \lambda J(\theta_1) + (1 - \lambda)J(\theta_2) \quad \text{for} \quad 0 < \lambda < 1 \tag{8.6}$$

If $J(\theta)$ is differentiable, it is convex if its Hessian, the matrix of second order derivatives with respect to θ, is positive definite.

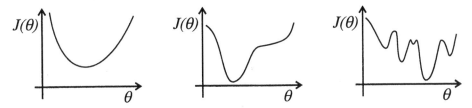

Figure 8.1 Convex, unimodal and nonunimodal $J(\theta)$

Remark A convex function is always unimodal. A unimodal function needs not be convex. ▮

Convex optimization problems satisfy these requirements. A problem is convex if both the cost function $J(\theta)$ and the allowable region, determined by the constraints $h(\theta)$, are convex. In general, it is not known whether $J(\theta)$ is convex or unimodal. Only for rather artificial cost functions, such as the sum of the squares of the prediction error if the error is a linear function of the parameters θ, can convexity of $J(\theta)$ be proved.

If neither convexity, nor unimodality can be assumed, a cost function $J(\theta)$ can have several minima. The minimum with the lowest value is called the global minimum, the other minima are called local minima. Once a local minimum has been detected, no statement whether the minimum is local or global can be made.

In Figure 8.1 the first cost function is convex (and hence unimodal), the second cost function is unimodal, and the third cost function is nonunimodal.

Depending on the function $J(\theta)$ and θ, a distinction among optimization problems can be made:

- If θ has integer values, an integer optimization problem arises. Examples of integer parameters are the order of a model and the size of a dead time if expressed as a number of times the sample time of a discrete model. If $J(\theta)$ is a nonlinear function of θ, no general solution technique exists. Only dynamic programming and enumeration, e.g., simulated annealing, may yield a solution. In general, these methods require large amounts of calculation time.

- If θ has real values, a solution can be found.

  - If $J(\theta)$ is linear, e.g., $J(\vartheta_1, \vartheta_2) = 3\vartheta_1 + 5\vartheta_2 - 6$, no minimum exists. Only if constraints are present, a minimum can be expected at a constraint.

– If $J(\theta)$ is quadratic, e.g., $J(\vartheta_1, \vartheta_2) = 2\vartheta_1^2 + 3\vartheta_2^2 + 5\vartheta_1\vartheta_2 - 14\vartheta_1 - 17\vartheta_2 + 10$, an easy solution can be obtained. Suppose $g(\theta)$ is the gradient, and hence the derivative of $J(\theta)$ with respect to θ. In a minimum θ^* $g(\theta^*) = 0$, for

$$J(\theta) > J(\theta^*) \qquad \forall \theta \qquad \text{with} \qquad \theta \neq \theta^* \qquad (8.7)$$

With $g(\theta^*) = 0$, the minimum θ^* is determined by the following equations:

$$g(\vartheta_1, \vartheta_2) = \begin{pmatrix} 4\vartheta_1 + 5\vartheta_2 - 14 \\ 5\vartheta_1 + 6\vartheta_2 - 17 \end{pmatrix} = \begin{pmatrix} 0 \\ 0 \end{pmatrix} \qquad (8.8)$$

which clearly yields the solution $\vartheta_1 = 1$ and $\vartheta_2 = 2$.

– If $J(\theta)$ is a nonlinear function of the parameter vector θ, then no general analytical solution is known. Apart from the fact that $g(\theta^*) = 0$, no other information is available. For this type of optimization problems the nonlinear optimization techniques are available. These techniques solve the optimization problem by searching for the optimum. Starting from an initial value of θ_0, a local search is made to determine an appropriate search direction d_0 in which a step s_0 is taken, so that

$$\theta_1 = \theta_0 + s_0 d_0 \qquad (8.9)$$

If θ_1 turns out to be successful, so that

$$J(\theta_1) < J(\theta_0) \qquad (8.10)$$

then θ_1 will be accepted as a new starting point for the next iteration. If θ_1 is not successful, so that

$$J(\theta_1) \geq J(\theta_0) \qquad (8.11)$$

then θ_1 will be rejected and, starting at θ_0, a new search direction d_0 and/or a new step size s_0 have to be selected.

This example illustrates that nonlinear optimization is an iterative procedure, starting at an initial guess θ_0 and terminating when no improvements can be found. Deciding whether or not improvements can be made depends on the selection of a stopping criterion. In a minimum the gradient $g(\theta) = 0$. So the value of $g(\theta)$ can be used for determining whether a minimum has been located. But also a test on the progress in decreasing $J(\theta)$ or the actual step size $\theta_{i+1} - \theta_i = s_i d_i$ can be used as a measure to stop the iterations. Suppose ε is a small positive constant. Then, the following criteria can be used as stopping criteria:

$$\| g(\theta) \| \leq \varepsilon \qquad (8.12)$$
$$|J(\theta_{i+1}) - J(\theta_i)| \leq \varepsilon \qquad (8.13)$$
$$\| \theta_{i+1} - \theta_i \| \leq \varepsilon \qquad (8.14)$$

8.2 Convergence

A very important characteristic of each optimization method is its ability to locate the minimum in a short space of time. Because optimization is an iterative process, the number of required iterations is an important measure for judging the convergence of an optimization method. Besides speed of convergence, robustness is also an important characteristic. An optimization method is robust if it is able to locate the minimum of almost any optimization problem in an acceptable number of iterations. There are optimization methods that are very fast for "smooth" functions but they are not able to locate the minimum of less smooth functions. Robustness is important because it makes methods reliable. Convergence is important because fast convergence makes the iterative process converge fast. Linear, super-linear and quadratic convergence can be distinguished.

A method has *linear convergence* if

$$0 < \beta = \frac{|\theta_{i+1} - \theta^*|}{|\theta_i - \theta^*|} < 1 \tag{8.15}$$

For any value of β where $0 < \beta < 1$, this method will converge to the optimum. The steepest descent turns out to have linear convergence. The value of β depends on the ratio r of the largest and smallest eigenvalue of the Hessian of $J(\theta)$, namely

$$\beta = \left(\frac{r-1}{r+1} \right) \tag{8.16}$$

This relation of β illustrates the importance of having a cost function in which all parameters are about equally important, so that the gradient $\partial J(\theta)/\partial \theta_i$ has about the same size for all i, $i = 1, \ldots, n$. If all parameters are about equally important, then all eigenvalues of the Hessian have about the same size and, consequently, r becomes about 1 and β becomes almost zero. With $\beta \approx 0$ the convergence will be very fast because then $\theta_{i+1} \approx \theta^*$. With $\beta = 0$ a method exhibits *super-linear convergence*.

A method has *quadratic convergence* if β^2 converges to a constant. The Newton method exhibits quadratic convergence.

Some methods are characterized by *quadratic termination*. This implies that these methods can accurately locate the minimum of a quadratic function $J(\theta)$ with n parameters θ_i, $i = 1, \ldots, n$, in a finite number of iterations. In general this number equals n.

Remark A method that has a fast convergence, is not necessarily fast. The total time needed for solving an optimization problem consists of the time needed for calculating $J(\theta_i)$ and $g(\theta_i)$ and the overhead of each method. Some methods have a small, others a large overhead. $\blacksquare$

8.3 Determination of step size s_i

As soon as a search direction d_i has been determined, a step s_i in the direction of d_i is taken, such that $\theta_{i+1} = \theta_i + s_i d_i$ is the minimum value in the direction d_i. So, s_i is the solution of the optimization problem

$$s_i = \arg \min_s J(\theta_i + s \cdot d_i) \tag{8.17}$$

Several methods can be used to determine s_i. They either search for the optimal value s_i, or they reduce an interval in which the minimum is located. Both approaches will be explained briefly.

Fixed and variable step

Searching with a fixed value of s_i is not attractive. Either this step is too small, which requires large amounts of calculation time, or the step is too large, which prevents the minimum to be found with a required accuracy. Consequently, variable step size methods must be introduced. As long as a step is successful, and $J(\theta_{i+1}) < J(\theta_i)$, the step size can be increased to a value larger than one, e.g., 1.3. As soon a failure has been detected, i.e., $J(\theta_{i+1}) > J(\theta_i)$, the step has to be decreased by multiplying it by a value smaller than 1, e.g., 0.213. The search process in the direction of d_i stops as soon as the step size falls below a predefined value.

Interpolation

A more systematic approach is to use some kind of interpolation. Suppose, three values of θ, such as θ_1, θ_2 and θ_3, are known together with their associated values of $J(\theta_i)$. These three points determine a second order polynomial in θ. The minimum of this polynomial θ' is an approximation of the minimum θ^* in the direction d_i. Select this point θ' instead of one of the three points θ_1, θ_2 and θ_3 and restart the interpolation process. In general, this interpolation quickly locates a minimum, especially for smooth functions. It is advisable first to locate an interval in which the minimum is located, in order to search for a set of parameters θ_1, θ_2 and θ_3, that satisfy $\theta_1 < \theta_2 < \theta_3$ and $J(\theta_1) > J(\theta_2)$ and $J(\theta_3) > J(\theta_2)$. This method is called *parabolic interpolation*.

If $J(\theta)$ is a quadratic function, then the minimum of $J(\theta)$ is calculated accurately in one step by this method. Higher order functions may require more iterations.

Golden section method

Instead of searching for the minimum or using interpolation, it is possible to reduce the uncertainty surrounding the location of the minimum by reducing the interval in which this minimum is located. This reduction can be done by the Golden section method. With each iteration this method reduces the interval by a fixed value k, with $k = 0.618034....$. This number k satisfies the relation $1/k = 1 + k$, which once was important in ancient Greek architecture. This special number would yield the highest visual satisfaction if it were used to determine the ratio between the height and width of buildings or illustrations. The Golden section method proceeds as follows. Select an initial interval $[\theta_1, \theta_2]$ with $\theta_1 < \theta^* < \theta_2$. Calculate two new points θ_3 and θ_4, with $0.5 < k < 1$:

$$\theta_3 = k.\theta_1 + (1-k).\theta_2 \tag{8.18}$$

$$\theta_4 = k.\theta_2 + (1-k).\theta_1 \tag{8.19}$$

Now, compare the values of $J(\theta_3)$ and $J(\theta_4)$.

- If $J(\theta_3) > J(\theta_4)$: $\theta_2 := \theta_2$; $\theta_1 := \theta_3$; $\theta_3 := \theta_4$ and θ_4 is calculated according to (8.19)

- If $J(\theta_3) < J(\theta_4)$: $\theta_1 := \theta_1$; $\theta_2 := \theta_4$; $\theta_4 := \theta_3$ and θ_3 is calculated according to (8.18)

Once an appropriate initial interval has been found, the Golden section method always converges to the minimum. This iterative procedure is continued until the interval $[\theta_1, \theta_2]$ has become less than a predefined small value ε.

Fibonacci method

The convergence speed of the Golden section method can be improved if the reduction factor k becomes iteration dependent. If this dependency is selected according to the Fibonacci sequence, the fastest interval reduction is achieved. The Fibonacci sequence is defined by

$$a_i = a_{i-1} + a_{i-2} \qquad (8.20)$$

with $a_0 = a_1 = 1$. This sequence looks like $1, 1, 2, 3, 5, 8, 13, 21, \ldots$. The reduction factor k_i in iteration i becomes

$$k_i = \frac{a_i}{a_{i+1}} \qquad (8.21)$$

For large values of i this ratio becomes the Golden section number $k_\infty = 0.618034\ldots$.

Comparison step-size methods

An optimization problem is solved with the aid of a number of line-search methods to determine the optimal value of s_i. The search direction d_i is calculated by the Fletcher-Powell method that is discussed in the next section.

$$\min_{\vartheta_1, \vartheta_2} \vartheta_1^2 + \vartheta_2^2 - \vartheta_1 \vartheta_2 \qquad (8.22)$$

with $\theta_0 = (-2, 8)^T$ and $s_0 = 0.1$.

The results, expressed as the number of times the function $J(\theta)$ is calculated, are illustrated in Table 8.1.

Table 8.1
Comparison line-search methods

Search method	Number of function evaluations
Fixed step	>200
Variable step	72
Golden section	63
Fibonacci	46
Parabolic interpolation	12

From this and many other examples, the following statements can be made: If the function $J(\theta_i)$ is smooth, interpolation yields faster results, while if $J(\theta_i)$ is not smooth the robust Golden section method or the Fibonacci method is to be preferred.

8.4 Determination of search direction d_i

The various nonlinear optimization methods can be categorized according to the way they determine an appropriate search direction d_i, as

- direct search methods

- gradient methods

- conjugate-gradient methods

As these names already indicate, the presence or absence of the gradient $g(\theta_i)$ is important. Direct search methods do not use gradient information while both other groups do utilize $g(\theta_i)$ in calculating d_i.

8.4.1   Direct search methods

Perpendicular search

Because no other information is available, perpendicular search uses the parameter axes as search directions d_i, so that $d_1 = (1,0,\ldots,0)^T, d_2 = (0,1,0,\ldots,0)^T$ and $d_n = (0,0,\ldots,0,1)^T$. In each direction d_i the minimum is sought by means of one of the one-dimensional optimization methods, e.g., interpolation or Golden section. If after n iterations θ_n does not satisfy the stop criterion, the whole process is repeated, by using $d_{n+1} = d_1, d_{n+2} = d_2$, etc. as illustrated in Figure 8.2.

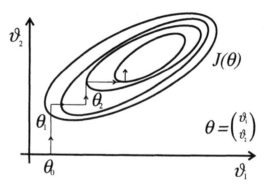

Figure 8.2 Perpendicular search

A disadvantage of perpendicular search is that the search directions cannot be adapted to the shape of the cost function $J(\theta)$. If the axes of the cost function do not coincide with the parameter axes, a slow convergence can be expected. This disadvantage is avoided in Powell's method.

Powell's method

Powell has proposed search directions that are adapted to the cost function. These directions turn out to be mutually conjugated to each other (see also Section 8.4.3) which indicates quadratic termination. An optimization problem with n parameters θ_i and with a quadratic cost function $J(\theta)$ can be solved by a maximum of n steps. In each step a one-dimensional optimization (line search) is solved. In Figure 8.3 the method is explained.

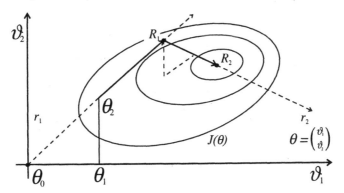

Figure 8.3 Powell's method

Starting with θ_0, the first n steps of perpendicular search are applied, and a search along the parameter axes is conducted. This yields θ_n. The direction $r_1 = \theta_n - \theta_0$ is a usable search direction, because it points in the direction in which the function has the largest decrease in value. Along this axis again a line search is made which yields the result R_1 of the first iteration. In the next iteration the last search direction d_n is replaced by the new search direction r_1, while the other directions d_i satisfy $d_i = d_{i+1}$, $i = 1, \ldots, n - 1$. After n line searches a new search direction is implemented, namely $r_2 = \theta_{2n} - R_1$. The minimum in this direction is called R_2. The new calculated search directions r_i are mutually conjugate (Section 8.4.3).

Consequently, each iteration consists of $n + 1$ line searches in the directions d_i, where the last direction is calculated on the basis of the results obtained. Powell's method adapts its search directions to the shape of the cost function $J(\theta)$. Its convergence is much better than the convergence of perpendicular search. Powell's method is one of the best known nonlinear optimization methods in cases where no gradient information is available.

Sometimes, the execution of the line search can be quite cumbersome. Then a search method without line search can be profitably applied. Two such methods exist, namely Pattern Search attributed to Hooke and Jeeves and the Simplex method attributed to Nelder and Mead.

Simplex method (Nelder and Mead)

The Simplex method utilizes mathematical figures to define a new point. No line search is required. In an n-dimensional space, each figure is defined by $n + 1$ points, so that there are triangles with 3 points for $n = 2$. A new point is calculated by mirroring the point with the highest function value against the average of all other points. If mirroring is successful, the size of the figure is expanded to improve convergence. If mirroring fails to locate a point with a lower function value, contraction is applied. As a consequence of this contraction, any required accuracy can be obtained. In Figure 8.4 mirroring, expansion and contraction are illustrated.

The Simplex method is comparatively reliable, because it can adapt the size and the shape of the figure to the cost function and line searches are not

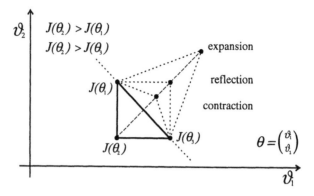

Figure 8.4 Simplex method: mirroring, expansion and contraction

required. The method is especially suited for nonsmooth cost functions with a few parameters. If many ($n > 10$) parameters have to be optimized, Powell's method and the gradient methods give a better performance.

8.4.2 Gradient methods

The negative gradient is the direction in which $J(\theta)$ shows, locally, its largest decrease. Consequently, a reasonable search direction d_i then becomes

$$d_i = -g(\theta_i) \tag{8.23}$$

If $J(\theta)$ is known analytically as a function of θ, the gradient can be calculated and used. If no analytical expression is available, the gradient $g(\theta) = ([g(\theta)]_1 \cdots [g(\theta)]_n)^T$ can be approximated with difference coefficients, as follows.

$$[g(\theta_i)]_j = \frac{J(\theta_i + \Delta n_j) - J(\theta_i)}{\Delta} \qquad j = 1, \ldots, n \tag{8.24}$$

or, more accurately but at the same time requiring more calculation time,

$$[g(\theta_i)]_j = \frac{J(\theta_i + \Delta n_j) - J(\theta_i - \Delta n_j)}{2\Delta} \qquad j = 1, \ldots, n \tag{8.25}$$

with Δ a small positive number and n_j the unit vector in direction j: $n_1 = (1, 0, 0 \cdots 0)^T$, $n_2 = (0, 1, 0 \cdots 0)^T$, etc.

With an analog computer the gradient method can be implemented because at each moment the search direction equals the negative gradient. With digital computers the steepest descent method is preferred. Each time a one-dimensional line search is conducted in the direction of the negative gradient. Once a line minimum θ_{i+1} is located, the gradient g_{i+1} is calculated and a new line search is made in the direction of the negative gradient. The steepest descent method exhibits linear convergence.

Gradient methods are more satisfactory than direct search methods if the gradient is available or if it can be calculated analytically. Even better convergence can be achieved by using conjugate-gradient methods.

8.4.3 Conjugate-gradient methods

Conjugate directions have attractive properties, because they have the quadratic termination characteristic. This states that the minimum of an n-dimensional quadratic function can be determined exactly in n steps if conjugate search directions are utilized. Powell's method yields conjugate search directions. In this section two other, gradient-based methods will be discussed with the same property.

Conjugate directions

The conjugate directions p_i for cost function $J(\theta)$ with

$$J(\theta) = \frac{1}{2}\theta^T A\theta + b^T\theta + c \tag{8.26}$$

satisfy

$$p_i^T A p_j = 0 \qquad \text{if } i \neq j; \qquad i = 1, \ldots, n; \ j = 1, \ldots, n \tag{8.27}$$
$$p_i^T A p_i = \Gamma_i \qquad\qquad\qquad i = 1, \ldots, n; \tag{8.28}$$

Suppose that search directions $p_j, j = 1, \ldots, i$, have been selected. Then

$$\theta_{i+1} = \theta_1 + s_1 p_1 + \cdots + s_k p_k + \cdots + s_i p_i = \theta_{k+1} + \sum_{j=k+1}^{i} s_j p_j \tag{8.29}$$

Then the gradient g_{i+1} in θ_{i+1} yields

$$g_{i+1} = A\theta_{i+1} + b = A\left(\theta_{k+1} + \sum_{j=k+1}^{i} s_j p_j\right) + b \tag{8.30}$$

so that

$$g_{i+1} = g_{k+1} + \sum_{j=k+1}^{i} s_j A p_j \tag{8.31}$$

Premultiplication with p_k^T yields

$$p_k^T g_{i+1} = p_k^T g_{k+1} + \sum_{j=k+1}^{i} s_j p_k^T A p_j \tag{8.32}$$

The first part of the right-hand side is zero, if in search direction p_k a minimum is detected by a line search. At a line minimum, the gradient g_{k+1} is perpendicular to p_k, so the inner vector product equals zero.

The second part of the right-hand side is zero if the search directions p_i are mutually conjugate for function $J(\theta)$. Consequently

$$p_k^T g_{i+1} = 0 \qquad \text{for} \qquad k = 1, 2, \ldots, i \tag{8.33}$$

After n iterations n linear independent search directions $p_i, i = 1, \ldots, n$, have been determined. Moreover, the inner vector product $p_k^T \cdot g_{n+1} = 0$ for $k = 1, \ldots, n$. These statements can only both be true if $g_{n+1} = 0$.

Consequently, if conjugate search directions are used AND in each search direction the minimum is determined exactly, the minimum of a quadratic cost function can be determined in exactly n steps. In the remaining part of this subsection two methods are presented that calculate conjugate directions, namely the methods of Fletcher and Reeves and the method of Fletcher and Powell.

Remark If the cost function $J(\theta)$ is not quadratic, the matrix A is approximated by the Hessian of $J(\theta)$.   ∎

Fletcher-Reeves

The conjugate directions are calculated according to

$$p_{i+1} = -g_{i+1} + \beta_i p_i \qquad (8.34)$$

with β_i, a scalar, satisfying

$$\beta_i = \frac{g_{i+1}^T g_{i+1}}{g_i^T g_i} \qquad (8.35)$$

The first search direction is selected to be the negative gradient, so $\beta_0 = 0$. This method is characterized by very simple calculations. Even if some more advanced methods require less iterations, this method can be competitive owing to its small overhead. In Figure 8.5 the calculation of p_{i+1} is elucidated.

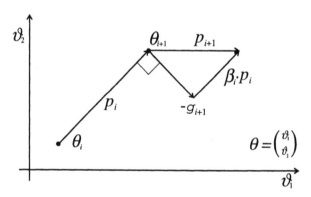

Figure 8.5 Calculation of p_{i+1} according to Fletcher-Reeves

If $J(\theta)$ is not quadratic, more than n steps are required to locate the minimum. It turns out to be advisable to reset the calculation of the conjugate search directions regularly, e.g., $\beta_{n+1} = \beta_{2n+1} = 0$. Otherwise, an accumulation of numerical errors and inaccuracies in determining the minimum of the one-dimensional line searches, may prevent the accurate calculation of conjugate directions.

Polak and Ribiere have proposed a modification in calculating β_i by introducing an automatic reset if not too much progress is made:

$$\beta_i = \frac{g_{i+1}^T(g_{i+1} - g_i)}{g_i^T g_i} \qquad (8.36)$$

Fletcher-Powell

The conjugate directions are calculated according to

$$p_i = -H_i g_i \qquad (8.37)$$

with

$$H_{i+1} = H_i - \frac{H_i \Delta g_i \Delta g_i^T H_i}{\Delta g_i^T H_i \Delta g_i} + \frac{\Delta \theta_i \Delta \theta_i^T}{\Delta \theta_i^T \Delta g_i} \qquad (8.38)$$

and

$$\Delta g_i = g_{i+1} - g_i \qquad (8.39)$$
$$\Delta \theta_i = \theta_{i+1} - \theta_i \qquad (8.40)$$

For $i = 0$, H_i is selected as a unity matrix, such that $p_0 = -g_0$. The matrix H_i will tend to the inverse Hessian of $J(\theta)$. This characteristic of the Fletcher-Powell method increases convergence. Knowledge of the Hessian A of the quadratic function $J(\theta) = \frac{1}{2}\theta^T A\theta + b^T\theta + c$, allows the calculation of the minimum θ^* of a quadratic function $J(\theta)$ in one step, namely

$$g(\theta) = A\theta + b \qquad (8.41)$$
$$g(\theta^*) = A\theta^* + b = 0 \qquad (8.42)$$

so that

$$\theta^* = \theta - A^{-1}g(\theta) \qquad (8.43)$$

This Newton algorithm demonstrates quadratic convergence. For quadratic functions $J(\theta)$ Fletcher and Powell proved that p_i, $i = 1, \ldots, n$ defines a set of conjugate directions and that $H_n = A^{-1}$. Consequently, Fletcher-Powell yields quadratic termination.

An important requirement for convergence is that H_i remains positive definite. Consequently, for real applications the positive definiteness of H_i has to be tested or H_i has to be reset at a regular interval. In practice, after n iterations the value of H_n is reset to the unity matrix I_n.

The Fletcher-Powell method is sometimes called a *variable metric method*, because the metric of the search process, as determined by H_i, changes depending on the shape of $J(\theta)$.

Compared with Fletcher-Reeves, the Fletcher-Powell method demonstrates a better convergence. Still, Fletcher-Reeves, or its modification Polak-Ribiere, is preferred for many applications where the overhead of Fletcher-Powell is too large. For example, in applications with large values of n (e.g., $n > 1000$), both the size of the matrix H, which requires n^2 memory locations, and the calculation time, can become too large. Then Fletcher-Reeves is preferred.

8.4.4 Example

To illustrate the behavior of several methods, the following quadratic function $J(x, y)$ will be minimized:

$$J(x, y) = x^2 - xy + y^2 \qquad (8.44)$$

with

$$g(x, y) = \begin{pmatrix} 2x - y \\ -x + 2y \end{pmatrix} = A \begin{pmatrix} x \\ y \end{pmatrix} \quad \text{and} \quad A = \begin{pmatrix} 2 & -1 \\ -1 & 2 \end{pmatrix} \quad (8.45)$$

Select $\theta_0 = (\begin{array}{cc} x_0 & y_0 \end{array})^T = (\begin{array}{cc} 1 & 2 \end{array})^T$.

Solution according to Fletcher-Reeves

$$p_0 = -g_0 = -g_0(1, 2) = \begin{pmatrix} 0 \\ 3 \end{pmatrix} \quad (8.46)$$

In this direction the minimum $\theta_1 = (\begin{array}{cc} 1 & 0.5 \end{array})^T$ can be found with one of the one-dimensional line search methods. The new search direction d_1 in θ_1 becomes, with $g_1(1, 0.5) = (\begin{array}{cc} 1.5 & 0 \end{array})^T$, according to Fletcher-Reeves

$$p_1 = -g_1 + \frac{g_1^T g_1}{g_0^T g_0} p_0 = \begin{pmatrix} -1.5 \\ 0 \end{pmatrix} + \frac{2.25}{9} \begin{pmatrix} 0 \\ -3 \end{pmatrix} = -\begin{pmatrix} 1.5 \\ 0.75 \end{pmatrix} \quad (8.47)$$

In this direction p_1 the minimum $\theta_2 = (\begin{array}{cc} 0 & 0 \end{array})^T$ can be located. In two steps (line minimizations) the minimum is accurately located, as illustrated in Figure 8.6, because Fletcher-Reeves exhibits the quadratic termination characteristic.

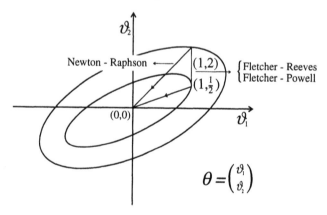

Figure 8.6 Search paths

Solution according to Fletcher-Powell
If H_0 is selected as the unity matrix, the first search direction p_0 becomes $p_0 = -g_0 = (\begin{array}{cc} 0 & -3 \end{array})^T$. In this direction the minimum $(\begin{array}{cc} 1 & 0.5 \end{array})^T$ is found, such that $g_1 = (\begin{array}{cc} 1.5 & 0 \end{array})^T$, $\Delta g_0 = g_1 - g_0 = (\begin{array}{cc} 1.5 & 3 \end{array})^T$ and $\Delta \theta_0 = \theta_1 - \theta_0 = (\begin{array}{cc} 0 & -1.5 \end{array})^T$.

According to the formulas of Fletcher-Powell (8.38), the new matrix H_1 becomes:

$$H_1 = H_0 - \begin{pmatrix} 0.2 & -0.4 \\ -0.4 & 0.8 \end{pmatrix} + \begin{pmatrix} 0.0 & 0.0 \\ 0.0 & 0.5 \end{pmatrix} = \begin{pmatrix} 0.8 & 0.4 \\ 0.4 & 0.7 \end{pmatrix} \quad (8.48)$$

The new search direction p_1 then becomes

$$p_1 = -H_1 g_1 = -\begin{pmatrix} 0.8 & 0.4 \\ 0.4 & 0.7 \end{pmatrix} \begin{pmatrix} 1.5 \\ 0.0 \end{pmatrix} = -\begin{pmatrix} 1.2 \\ 0.6 \end{pmatrix} \quad (8.49)$$

In this direction again the minimum $\theta_2 = (\ 0 \quad 0\ )^T$ is found. According to the theory, H_2 has to become the inverse Hessian of $J(\theta)$. With $\Delta\theta_1 = \theta_2 - \theta_1$ and $\Delta g_1 = g_2 - g_1 = (\ -1.5 \quad 0\ )^T$, H_2 becomes:

$$H_2 = H_1 - \begin{pmatrix} 0.8 & 0.4 \\ 0.4 & 0.2 \end{pmatrix} + \begin{pmatrix} 2/3 & 1/3 \\ 1/3 & 1/6 \end{pmatrix} = \begin{pmatrix} 2/3 & 1/3 \\ 1/3 & 2/3 \end{pmatrix} \tag{8.50}$$

This value of H_2 satisfies $H_2.A^{-1} = I$, so indeed H_2 is the inverse Hessian of $J(\theta)$.

Exercise
Calculate the value of θ_2 by using the gradient method or perpendicular search.

8.4.5 Comparison search methods

Two optimization problems are solved with the aid of a number of search methods that determine a value of d_i. The step size s_i is calculated with parabolic interpolation. The gradient $g(\theta)$ is obtained by forward difference coefficients.

Problem 1:

$$\min_{\vartheta_1,\vartheta_2} \vartheta_1^2 + \vartheta_2^2 - \vartheta_1\vartheta_2 \tag{8.51}$$

with $\theta_0 = (-2,8)^T$ and a stop criterion $\| g(\theta) \| \leq 0.03$.

Problem 2:

$$\min_{\vartheta_1,\vartheta_2,\vartheta_3,\vartheta_4} \{100 \cdot (\vartheta_1^2 - \vartheta_2)^2 + (1 - \vartheta_1)^2 + 90 \cdot (\vartheta_3^2 - \vartheta_4)^2 + (1 - \vartheta_3)^2$$

$$+10.1 \cdot [(\vartheta_2 - 1)^2 + (\vartheta_4 - 1)^2] + 19.8 \cdot (\vartheta_2 - 1)(\vartheta_4 - 1)\} \tag{8.52}$$

with $\theta_0 = (-2,8,-2,8)^T$ and a stop criterion $\| g(\theta) \| \leq 0.03$.
The results, expressed as the number of times NR (NR_1 for problem 1 and NR_2 for problem 2) the function $J(\theta)$ is calculated, are illustrated in Table 8.2.

Table 8.2
Comparison of search methods

Search method	NR_1	NR_2
Perpendicular search	70	496
Powell	35	287
Simplex	66	244
Steepest descent	49	>500
Fletcher-Reeves	21	132
Fletcher-Powell	21	129

From this and many other examples, the following statements can be made. If the function $J(\theta)$ is smooth, the conjugate-gradient methods yield fast results, while if $J(\theta)$ is not smooth the robust Simplex or pattern search methods have to be preferred. Both perpendicular search and steepest descent need too many function evaluations. Owing to the slightly better convergence of Fletcher-Powell, this method has some advantages over Fletcher-Reeves. Only if the number of parameters becomes large, does Fletcher-Reeves have a clear advantage over Fletcher-Powell, because it does not need matrices for the calculation of the conjugate directions.

8.5 *Least Squares Problem*

An important application area of optimization is identification or parameter estimation. In general, the sum of the squares of the error $\epsilon(k)$ is taken, with $\epsilon(k, \theta)$ the error between measurements $\{y(k)\}_N$ and the predicted value of the model response $\{y_m(k, \theta)\}_N$.

$$J_N(\theta) = \frac{1}{2N} \sum_{k=1}^{N} \epsilon^2(k, \theta) = \frac{1}{2N} \sum_{k=1}^{N} [y(k) - y_m(k, \theta)]^2 \qquad (8.53)$$

If only $J_N(\theta)$ is known, a general nonlinear optimization problem arises that can be solved by any of the methods dealt with in the previous section. So, the same characteristics can be distinguished. In general, there is no guarantee of convergence, and a time-consuming, iterative procedure is needed to find a solution. However, if not only $J_N(\theta)$ but also the terms $\epsilon(k, \theta)$ are known, optimization methods are available that yield a better convergence. In this section these methods are treated.

Define $E(\theta) = \begin{pmatrix} \epsilon(1|\theta) & \cdots & \epsilon(N|\theta) \end{pmatrix}^T$ and $\theta = (\vartheta_1, \ldots, \vartheta_n)^T$. Hence, $J(\theta) = E^T E / 2N$. The elements $\Phi_{jk}(\theta)$ of the $n \times N$ matrix $\Phi(\theta)$ are defined as

$$\Phi_{jk}(\theta) = \frac{\partial \epsilon(k|\theta)}{\partial \vartheta_j} \qquad (8.54)$$

Then the gradient $g(\theta)$ of $J_N(\theta)$ (8.53) is

$$g(\theta) = \frac{\partial J(\theta)}{\partial \theta} = \begin{pmatrix} \frac{\partial J}{\partial \vartheta_1} & \cdots & \frac{\partial J_N}{\partial \vartheta_n} \end{pmatrix}^T = \frac{1}{N} \Phi(\theta) E(\theta) \qquad (8.55)$$

The Hessian H or matrix of second order derivatives of $J_N(\theta)$ yields

$$H = \frac{\partial^2 J_N}{\partial \theta^2} = \frac{1}{N} \Phi(\theta) \Phi(\theta)^T + \frac{1}{N} \frac{\partial \Phi(\theta)}{\partial \theta} E(\theta) \qquad (8.56)$$

Now a distinction has to be made between ARX model structures and the other models structures such as ARMAX, OE and BJ.

- An ARX model is defined by

$$A(q)y(k) = B(q)u(k) + e(k) \qquad (8.57)$$

with prediction error $\epsilon(k|\theta)$

$$\epsilon(k|\theta) = \hat{y}(k) - y(k) = \theta^T \phi(k) - y(k) \qquad (8.58)$$

The prediction error $\epsilon(k|\theta)$ of an ARX model is linear in the parameter vector θ. Consequently, $\Phi(\theta)$ does not depend on θ and is constant. The Hessian H becomes $H = \frac{1}{N} \Phi \Phi^T$ and H is independent from θ.

- An ARMAX model is defined by

$$A(q)y(k) = B(q)u(k) + C(q)e(k) \qquad (8.59)$$

with prediction error $\epsilon(k|\theta)$

$$\epsilon(k|\theta) = \hat{y}(k) - y(k) = \theta^T \phi(k, \theta) - y(k) \qquad (8.60)$$

The prediction error $\epsilon(k|\theta)$ of an ARMAX model is not linear in the parameter vector θ, because $\phi(k|\theta)$ depends not only on the measured data $\{y(k), u(k)\}_N$, but also on θ. Consequently, $\Phi(\theta)$ is still a function of θ. The Hessian $H(\theta)$ is a function of θ and is defined by (8.56). This equation is rather difficult to calculate.

In solving the least square problem with knowledge of both the gradient $g(\theta)$ and Hessian $H(\theta)$, the Newton algorithm is preferred:

$$\theta_{i+1} = \theta_i - H^{-1}(\theta)g_i(\theta) \qquad (8.61)$$

Because sometimes the value of H is not accurately known, a step size s_i is introduced to avoid excessively large steps.

Now four different selections of H are discussed

- Select H according to (8.56). The ARX model structure yields a simple calculation of H. In one step the Newton method calculates the required optimal value θ^*, namely

$$\theta^* = \theta_i - (\Phi\Phi^T)^{-1}\Phi E(\theta_i) \qquad (8.62)$$

Any value of θ_i can be used. By taking $E(\theta_i) = \Phi^T\theta_i - Y$ with $Y = (y(1)\ldots y(N)^T$ (8.62) becomes

$$\theta^* = \theta_i - (\Phi\Phi^T)^{-1}\Phi(\Phi^T\theta_i - Y) = (\Phi\Phi^T)^{-1}\Phi Y \qquad (8.63)$$

which are the *Normal equations*. Consequently, the minimization of the cost function of an ARX model structure, is achieved in one step by the application of the Newton method. So, the solution is calculated analytically.

- Approximate H by $H \approx \Phi\Phi^T/N$. If Φ is a function of θ, (8.56) cannot easily be calculated. Then this approximation of H yields the Gauss-Newton method for ARMAX model structures, so that

$$\theta_{i+1} = \theta_i - s_i H^{-1}g(\theta_i)$$
$$\approx \theta_i - s_i \left(\frac{1}{N}\Phi\Phi^T\right)^{-1} \frac{1}{N}\Phi E(\theta_i)$$
$$= \theta_i - s_i \left(\Phi\Phi^T\right)^{-1} \Phi E(\theta_i) \qquad (8.64)$$

It takes several or many iterations for θ_i to converge to θ^*. Solving the minimization problem with an ARMAX, but also with OE or BJ model structures, is a time consuming activity as a consequence of the iterative solution by nonlinear optimization. In the vicinity of the minimum this algorithm converges very fast, because then $J_N(\theta)$ can be approximated by means of a quadratic function in θ. However, if $J(\theta)$ is not quadratic in θ, the approximation $H \approx \Phi\Phi^T/N$ is not true and, consequently, convergence can not be guaranteed. To overcome these convergence problems, Levenberg and Marquardt have proposed a modification to the Gauss-Newton algorithm. With the aid of the parameter λ the character of the Levenberg-Marquardt algorithm can be modified.

- The Levenberg-Marquardt method

$$\theta_{i+1} = \theta_i - s_i(\lambda I + H)^{-1}g(\theta_i) \approx \theta_i - s_i\left(\lambda I + \frac{1}{N}\Phi\Phi^T\right)^{-1}\frac{1}{N}\Phi E(\theta_i) \quad (8.65)$$

is a more reliable method. If $\lambda=0$, the original Gauss-Newton method is retrieved. With λ large, the algorithm tends to become a steepest descent algorithm.

- Select $H = I$, so that search directions according to the steepest descent method are taken.

$$\theta_{i+1} = \theta_i - s_i.g_i(\theta) \approx \theta_i - s_i\frac{1}{N}\Phi E(\theta_i) \quad\quad (8.66)$$

The proposed methods of Gauss-Newton and its modification attributed to Levenberg and Marquardt yield a better result than a general purpose minimization method. The difference can be ascribed to knowledge of $\epsilon(k|\theta)$ and an analytical expression of the gradient $g(\theta)$ instead of only knowledge of $J_N(\theta)$.

8.6 *Optimization with constraints*

Up to now problems have been discussed for solving optimization problems without constraints. In this section the solution of constrained optimization problems will be discussed. These constraints are represented as $h_j(\theta)$, $j = 1,\ldots,m$.

In general, these constraints are formulated as inequality constraints. In some optimization problems with m constraints, a maximum of n constraints is formulated as equality constraints. So, the general constrained optimization problem becomes

$$\theta^* = \arg\min_\theta J(\theta) \quad\quad (8.67)$$
$$h_j(\theta) = 0 \quad j = 1,\ldots,m_1$$
$$h_j(\theta) \leq 0 \quad j = m_1+1,\ldots,m$$

If $m_1 = n$, a set of n equality constraints with n unknowns arises. The solution of this set of equations provides the solution of the optimization problem. Consequently, no search is required. The equality constraints completely determine the solution. If $m_1 > n$, no solution is possible. If $m_1 < n$, a solution can be obtained by one of the following approaches

- Elimination

- Lagrange multipliers

- Gradient-projection method

- Reduced-gradient method

- Penalty functions

In the next part, two elegant methods will be discussed that allow the presence of inequality constraints in problem formulation, namely the gradient-projection method and the reduced-gradient method. These methods use a set of active equality constraints. As soon as the search path attempts to leave the allowable region as defined by the inequality constraints, the activated inequality constraint is added to the set of already active equality constraints. Consequently, $m_1 := m_1 + 1$. If an inequality constraint is not active, it is neglected in calculating an appropriate search direction. Also the set of active equality constraints can be reduced if the search direction points into the allowable region, so m_1 reduces to $m_1 := m_1 - 1$.

8.6.1 Elimination

There are several approaches for solving the constrained optimization problem. The methods try to transform the original constrained problem into some kind of unconstrained problem. For example, if θ has to be positive, it can be replaced by another parameter, e.g., x, with $\theta = x^2$. In the cost function J, θ is replaced by x^2 and the optimization can start to solve the new unconstrained problem $\min J(x^2)$ for x. This substitution guarantees that θ will always be positive.

If a parameter has both a lower and an upper bound, then this parameter can be replaced by $a.\sin(x) + b$. This function is always limited between $b - a$ and $b + a$.

Linear equality constraints can easily be removed. Suppose that there are m_1 equality constraints with $m_1 < n$. Now the parameter vector θ can be separated into an $m_1 \times 1$ vector z and an $(n - m_1) \times 1$ vector y

$$\theta = \begin{pmatrix} y \\ z \end{pmatrix} \tag{8.68}$$

Suppose that the equality constraints can be written as $K\theta = b$. With $K = [\ K_1 \quad K_2\ ]$ with K_1 an $m_1 \times (n - m_1)$ matrix and K_2 a square $m_1 \times m_1$ matrix the equality constraints can be expressed as

$$K\theta = K_1 y + K_2 z = b \tag{8.69}$$

If the set of equality constraints is independent, the inverse of K_2 exists, and hence

$$z = K_2^{-1}(b - K_1 y) \tag{8.70}$$

By substitution of the $m_1 \times 1$-vector z by the $(n - m_1) \times 1$-vector y, the equality constraints have been removed in the problem formulation. Moreover, the dimension of the unknown parameter vector θ has been reduced to the dimension $(n - m_1)$.

8.6.2   *Lagrange multipliers*

The constraint optimization problem with m equality constraints, such as

$$\theta^* = \arg \min_{\theta} \quad J(\theta) \tag{8.71}$$
$$h(\theta) = 0$$

with θ an n-dimensional and h an m-dimensional vector, can be reformulated as an unconstrained optimization problem by introducing m Lagrange multipliers λ in

$$(\theta^*, \lambda^*) = \arg \min_{\theta, \lambda} J(\theta, \lambda) + \lambda^T h(\theta) \tag{8.72}$$

The solution of the unconstrained problem is defined by $n + m$ equations of the gradient of the Lagrange function $L(\theta, \lambda) = J(\theta, \lambda) + \lambda^T h(\theta)$ for θ and λ. This gradient is zero. These $n + m$ equations are necessary and sufficient for solving the n (θ) plus m (λ) unknowns.

Example

$$\min_{\vartheta_1, \vartheta_2} \vartheta_1^2 + \vartheta_2^2 \vartheta_1 + \vartheta_2 = 1 \tag{8.73}$$

The Lagrange function becomes $L(\vartheta_1, \vartheta_2, \lambda) = \vartheta_1^2 + \vartheta_2^2 + \lambda(\vartheta_1 + \vartheta_2 - 1)$. The partial derivatives of $L(\vartheta_1, \vartheta_2, \lambda)$ yield

$$\frac{\partial L(\vartheta_1, \vartheta_2, \lambda)}{\partial \vartheta_1} = 2\vartheta_1 + \lambda = 0 \tag{8.74a}$$

$$\frac{\partial L(\vartheta_1, \vartheta_2, \lambda)}{\partial \vartheta_2} = 2\vartheta_2 + \lambda = 0 \tag{8.74b}$$

$$\frac{\partial L(\vartheta_1, \vartheta_2, \lambda)}{\partial \lambda} = \vartheta_1 + \vartheta_2 - 1 = 0 \tag{8.74c}$$

These three linear equations define the solution of the three unknowns, namely $\vartheta_1 = \vartheta_2 = \frac{1}{2}$ and $\lambda = -1$.

8.6.3   *Gradient-projection method*

The gradient-projection method is based on the steepest descent method for unconstrained optimization problems, as illustrated in Figure 8.7.

Suppose that there are m_1 equality constraints and $(m - m_1)$ inequality constraints. Then the equality constraints can be written as

$$K\theta = b \qquad K \; : \; m_1 \times n \tag{8.75}$$

The projection matrix P projects the gradient $g(\theta)$ on all active equality constraints. So, a search along the projected gradient ensures that the equality constraints are satisfied. P is defined as

$$P = I - K^T(KK^T)^{-1}K \tag{8.76}$$

The search direction d_i is the projected negative gradient, formulated as

$$d_i = -Pg_i \tag{8.77}$$

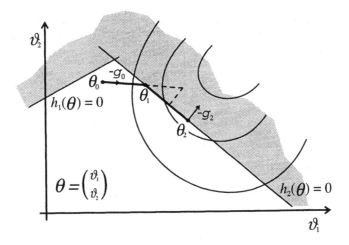

Figure 8.7 Gradient-projection method

This projection method, as proposed by Rosen (1960), can also be utilized for the Fletcher-Reeves conjugate gradient method (Van den Bosch and Lootsma, 1987).

The calculation of the projection matrix P is a nice example of the power of optimization, because the derivation of P can be formulated as an optimization problem as follows.

The search direction d_i has to satisfy the following requirements

- d_i has to decrease the cost function J as much as possible;

- d_i has to satisfy all active equality constraints;

- d_i has to be normalized.

These requirements can be formulated as an optimization problem. The first requirement can be implemented by selecting $J = d_i^T.g_i$. For, the best search direction is the negative gradient. If d_i is the negative gradient, J, the inner product of the two vectors, will receive its lowest value.

The second requirement can be formulated as $Kd_i = 0$, with K the description of the active linear or linearized equality constraints. Once θ_{i+1} satisfies $K\theta_{i+1} = b$, the search path has to satisfy $Kd_i = 0$.

The third requirement is introduced to avoid infinitely large values of d_i. An arbitrary, but useful choice of normalization is $d_i^T.d_i = 1$. Consequently, the solution of the following optimization problem will yield an optimal value of the search direction d_i

$$\min_{d_i} d_i^T g_i \tag{8.78}$$

with constraints

$$Kd_i = 0 \tag{8.79a}$$

$$d_i^T d_i - 1 = 0 \tag{8.79b}$$

The analytical solution of this optimization problem can be obtained by introducing Lagrange multipliers λ and β, with λ an $m_1 \times 1$ vector and β a scalar. The Lagrangian function L is stated as

$$L(d_i, \lambda, \beta) = d_i^T g_i + \lambda^T K d_i + \beta(d_i^T d_i - 1) \qquad (8.80)$$

The solution satisfies

$$\frac{\partial L(d_i, \lambda, \beta)}{\partial d_i} = 0; \qquad \frac{\partial L(d_i, \lambda, \beta)}{\partial \lambda} = 0; \qquad \frac{\partial L(d_i, \lambda, \beta)}{\partial \beta} = 0 \qquad (8.81)$$

The partial derivative of L to d_i yields

$$g_i + K^T \lambda + 2\beta d_i = 0 \qquad (8.82)$$

So,

$$d_i = \frac{-1}{2\beta}(g_i + K^T \lambda) \qquad (8.83)$$

Using this value of d_i in $K.d_i = 0$ and solving for λ results in

$$\lambda = (KK^T)^{-1} K g_i \qquad (8.84)$$

and

$$d_i = \frac{-1}{2\beta}(I - K^T (KK^T)^{-1} K) g_i \qquad (8.85)$$

This "optimal" search direction d_i coincides with the previously calculated projected gradient. Only the scaling factor $1/2\beta$, which has no influence on the search direction, makes some difference.

8.6.4   Reduced-gradient method

Suppose a set of m inequality constraints is present in formulation of the problem. At point θ_i a set of m_1 active linear or linearized equality constraints is detected, namely $K\theta = b$ with K an $m_1 \times n$ matrix. The parameter vector θ can be separated into the m_1 vector z and the $(n - m_1)$ vector y, so that

$$K_1 y + K_2 z = b \qquad (8.86)$$

The cost function $J(\theta)$ in the $n \times 1$ parameter vector θ can be reduced to the cost function $F(y)$ in the $(n - m_1)$ parameter vector y, namely

$$F(y) = J\left[y, K_2^{-1}(b - K_1 y)\right] \qquad (8.87)$$

The gradient $g_F(y)$ of $F(y)$ with respect to y, is defined as

$$g_F(y) = \nabla_y J(y) - \nabla_z J(z) K_2^{-1} K_1 \qquad (8.88)$$

By selecting the search direction $d_i(y) = -g_F(y)$ it is assured that at any time the equality constraints $K\theta = b$ are satisfied. In an $(n - m_1)$ dimensional parameter space a search is made for optimal values of y by minimizing $F(y)$. The solution y^* yields $z^* = K_2 P^{-1} K_1 y^*$ and so $\theta^* = \begin{pmatrix} y^{*T} & z^{*T} \end{pmatrix}^T$.

   The reduced-gradient method, like the gradient-projection method, can be profitably used for solving nonlinear optimization problems with linear constraints. If these constraints are nonlinear, they have to be linearized in θ_i or penalty functions have to be introduced.

8.6.5 Penalty functions

Both previously mentioned methods (the gradient-projection method and the reduced-gradient method) use a set of active equality constraints. Another approach removes all constraints in the constrained optimization problem by adding them, via a penalty function J_p, to the cost function J which yields a new cost function $J_m = J + J_p$. The penalty function J_p will have a low value, preferably zero, in the allowable region and a large value outside this area, as illustrated in Figure 8.8.

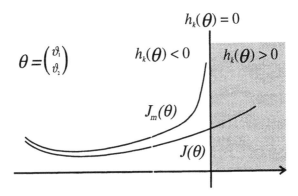

Figure 8.8 Cost function $J(\theta)$ and modified cost function $J_m(\theta)$

The penalty function J_p takes care that during the search process any attempt to leave the allowable region $(h_j(\theta) < 0)$ will be prevented by the added large value of J_p to J. So, by using penalty functions the original constrained optimization problem

$$\min_{\theta} \quad J(\theta) \tag{8.89}$$
$$h_j(\theta) \quad \le \quad 0$$

is replaced by the following unconstrained optimization problem

$$\min_{\theta} J(\theta) + \alpha J_p(\theta) \tag{8.90}$$

For α small, the constraints do not play an important part in J_m, while for increasing values of α the constraints become dominant as soon as the search path approaches a constraint. Too large a value of α has to be avoided because in the neighborhood of the inequality constraints $h_j(\theta)$ the gradient changes rapidly, which hampers convergence.

Examples of penalty functions are

- Internal penalty functions (the search path never crosses the inequalities)

$$J_p(\theta) = \sum_{j=1}^{m} \ln(-h_j(\theta)) \tag{8.91}$$

$$J_p(\theta) = \sum_{j=1}^{m} \frac{1}{h_j(\theta)} \tag{8.92}$$

- External penalty functions (the search path can cross the inequalities)

$$J_p(\theta) = \sum_{j=1}^{m} \max(0, h_j(\theta)) \tag{8.93}$$

$$J_p(\theta) = \sum_{j=1}^{m} \max^2(0, h_j(\theta)) \tag{8.94}$$

Note that the penalty function should be continuous and differentiable if gradient-based methods are used for solving the associated unconstrained optimization problem.

8.7 Summary

We have shown that optimization is a general tool for system analysis and system design. Methods are discussed that enable the solution of a general optimization problem

$$\theta^* = \min_{\theta} J(\theta) \tag{8.95}$$

Nonlinear optimization problems are solved by nonlinear optimization methods. These iterative methods search in the parameter space θ for values of θ_{i+1} which criterion values $J(\theta_{i+1})$ have a lower value than the present value $J(\theta_i)$ in θ_i. Only for convex and unimodal functions $J(\theta)$ can convergence be guaranteed by this search process. If the minimum of a nonunimodal function is searched for, only a local minimum is found. No statement can be made whether this minimum is also the global minimum of this function.

This iterative search process consists of two steps. First, a search direction d_i is calculated and next a step s_i is made in the direction of d_i yielding the new exploratory point $\theta_{i+1} = \theta_i + s_i d_i$. Depending on prior knowledge of the criterion $J(\theta)$, suitable methods can be selected. The smoother $J(\theta)$ is, the more accurate and faster are the methods that can be selected. Especially, the use of gradient information of $J(\theta)$ will improve convergence. If even the Hessian H of J is known, the very fast Newton iterative scheme is available.

We have shown that the structure of a least square problem, which arises in many parameter estimation problems, can be exploited for obtaining fast and accurate solutions. For ARX models, with a prediction error linear in the parameter θ, the solution is calculated analytically. No iterations are required. In using the other model structures, such as ARMAX, Output Error or Box-Jenkins, the prediction error is a nonlinear function of θ. Then, a time-consuming iterative search is needed to locate the optimal parameters, namely those parameters that yield the lowest value of the criterion.

Optimization problems with constraints on the parameters θ require dedicated search algorithms. Either the constraints are eliminated (reduced-gradient method, direct elimination), or included in the criterion by means of a penalty function or used as a projection for the search direction (gradient-projection method).

8.8 References

Solving nonlinear optimization problems is a basic part of numerical analysis.
Many books discuss with some depth these optimization problems.

We will especially mention some books fully devoted to optimization,
namely Adby and Dempster (1974), Rao (1978), Schwefel (1981) and Dennis
and Schnabel (1983).

Leigh (1983) contains a short, but valuable contribution of nonlinear optimization methods.

8.9 Problems

1. Solve the following optimization problem

$$\min_{x,y} \quad x^2 + y^2 + xy$$

$$
\begin{array}{rcl}
x + y & \geq & 5 \\
x & \leq & 5 \\
y & \leq & 3
\end{array}
$$

   (a) Calculate the gradient $g(x,y)$ in $(x,y)=(3,2)$

   (b) Calculate the search direction of the steepest descent method in (3,2),
       if the constraints are neglected.

   (c) Calculate the search direction of perpendicular search in (3,2), if the
       constraints are neglected.

   (d) Calculate the search direction of the gradient-projection method in
       (3,2) if all 3 constraints are present.

2. Solve the following optimization problem

$$\min_{x,y} J(x,y)$$

   with
$$J(x,y) = 2x^2 + y^2 + xy$$

   (a) Calculate the gradient $g(x,y)$ and the Hessian H (matrix of second
       derivatives) of $J(x,y)$.

   (b) Calculate the minimum of $J(x,y)$ with the Newton method.

   (c) Calculate the first search direction p_o in $\theta_o = (0,1)^T$ of the Fletcher-
       Reeves method.

   (d) After 1 iteration a line minimum is found, namely $\theta_1 = (\frac{-5}{16}, \frac{3}{8})^T$.
       Show that θ_1 is really the line minimum in the direction of p_o.

   (e) Calculate in θ_1 the new search direction p_1.

   (f) Show that p_o and p_1 are conjugate directions with respect to $J(x,y)$.

3. Minimize the following criterion function $J(x,y) = x^2 + y^2 + xy + 5$.

   (a) Calculate (analytically) the gradient g and Hessian H of $J(x,y)$.

(b) Calculate the minimum of $J(x, y)$ with the Newton method.

(c) Determine the first search direction p_0 according to Fletcher-Reeves, starting with (1,0). In this direction, a line minimum is detected, namely $(\frac{4}{14}, \frac{-5}{14})$.

(d) Show that this point is indeed the minimum in the direction p_0.

(e) Calculate in this line minimum the new search direction p_1.

(f) Show that p_0 and p_1 are conjugate search directions of $J(x, y)$.

4. Proof that the minimum of a quadratic function

$$J(\theta) = \frac{1}{2}\theta^T A\theta + b\theta + c$$

with n parameters can be found in n steps. Which requirements are imposed to obtain this result?

Literature

Adby, P.R. and Dempster, M.A.H. (1974): *Introduction to Optimization Methods*, Chapman and Hall, London.

Akaike, H. (1969): "Fitting autoregressive models for prediction", *Ann. Inst. Math.*, **21**, 243–347.

Akaike, H. (1974): "A new look at the statistical model identification", *IEEE Trans. Autom. Contr.*, **AC-19**, 716–723.

Akaike, H. (1981): "Modern development of statistical methods", in *Trends and progress in system identification* (P. Eykhoff ed.), Pergamon Press, 169–184.

Åström, K.J. and Wittenmark, B. (1984): *Computer Controlled Systems*, Prentice-Hall, Englewood Cliffs, New Jersey.

Backx, T. (1987): *Identification of an Industrial Process: a Markov Parameter Approach*, PhD dissertation, Eindhoven University of Technology.

Beck, J.V. and Arnold, K.J. (1977): *Parameter Estimation in Engineering and Science*, John Wiley & Sons, New York.

Brenan, K.E., Campbell, S.L. and Petzold, L.R. (1989): *Numerical Solution of Initial-Value Problems in Differential-Algebraic Equations*, North-Holland, New York.

Caines, P.E. (1988): *Linear Stochastic Systems*, Wiley series in probability and mathematical statistics, John Wiley & Sons, New York.

Cellier, F.E. (1990): *Continuous System Modeling*, Springer Verlag, Berlin.

Dennis, J.E. and Schnabel, R.B. (1983): *Numerical methods for Unconstrained Optimization of Nonlinear Equations*, Prentice Hall, Englewood Cliffs, New Jersey.

Eykhoff, P. (1974): *System Identification*, John Wiley, London.

Federov, J.J. (1972): *Theory of Optimal Experiments*, Academic Press, New York.

Gear, C.W. (1971): *Numerical Initial Value Problems in Ordinary Differential Equations*, Prentice-Hall, Englewood Cliffs, New Jersey.

Goodwin, G.C., and Payne, R.L. (1977): *Dynamic System Identification: Experiment Design and Data Analysis*, Academic Press, New York.

Graupe, D. (1972): *Identification of Systems*, Van Nostrand Reinhold Company, New York.

Hairer, E.H., Lubich, C. and Roche, M. (1989): *The Numerical Solution of Differential Algebraic Systems by Runge-Kutta methods*, Lecture notes in mathematics, No 1409, Springer Verlag, Berlin.

Hannan, E.J. and Deistler, M. (1988): *The Statistical Theory of Linear Systems*, Wiley series in probability and mathematical statistics, John Wiley & Sons, New York.

Jenkins, G.M. and Watts, D.G. (1969): *Spectral Analysis and its Applications*, Holden-Day, San Francisco, California.

Kailath, T. (1980): *Linear Systems*, Prentice-Hall International, UK.

Karnopp, D.C., Margolis, D.L. and Rosenberg, R.C. (1990): *System Dynamics, a Unified Approach*, John Wiley & Sons, New York.

Kheir (ed.), N.A. (1988): *Systems Modeling and Computer Simulation*, Marcel Dekker Inc., New York.

Kwakernaak, H. and Sivan, R. (1991): *Modern Signals and Systems*, Prentice-Hall, Englewood Cliffs, New Jersey.

Leigh, J.R. (1983): *Modelling and Simulation*, IEE Topics in Control Engineering, Peter Peregrinus Ltd, London.

Ljung, L., and Söderström, T. (1983): *Theory and Practice of Recursive Identification*, The MIT Press, Cambridge, Massachusetts.

Ljung, L. (1987): *System Identification: Theory for the User*, Prentice-Hall, Englewood Cliffs, New Jersey.

Ljung, L. (1991): *System Identification Toolbox – User's Guide*, For use with Matlab.

Mehra, R.K. (1974): "Optimal input signals for parameter estimation in dynamic systems, survey and new results", *IEEE Trans. Autom. Contr.*, **AC-19**, 753–768.

Mehra, R.K. and Lainiotis, D.G. (1976): *System Identification – Advances and Case Studies*, Academic Press, New York.

Mehra, R.K. (1981): Choice of input signals, in *Trends and Progress in System Identification*, edited by P. Eykhoff, Pergamon Press, 305–366.

Nelder, J.A. and Mead, R. (1965): "A simplex method for function minimization", *Computer Journal*, **7**, 308-313.

Priestley, M.B. (1989): *Spectral Analysis and Time Series, Vol. 1, Univariate series*, Academic Press, London.

Rao, S.S. (1978): *Optimization, Theory and Applications*, Wiley Eastern Limited, New Delhi.

Richalet, J., Rault, A. and Pouliquen, R. (1971): *Identification des Processus par la Méthode du Modèle*, Gordon and Breach, Paris (in French).

Richalet, J. (1991): *Pratique de l'Identification*, Editions Hermès, Paris (in French).

Rosen, J.B. (1960): *The gradient projection method for non-linear programming, Linear Constraints.* Journal SIAM, **8**, 181-217.

Sargent, R.G. (1988): "A tutorial on validation of simulation models", *Proc. 1988 Winter Simulation Conf.*

Schoukens, J. and Pintelon, R. (1991): *Identification of Linear Systems – a Practical Guideline to Accurate Modeling,* Pergamon Press, Oxford.

Schroeder, M.R. (1970): "Synthesis of low peak factor signals and binary sequences with low autocorrelation", *IEEE Trans. Inform. Theory,* **IT-16**, 85–89.

Schwefel, H.P. (1981): *Numerical Optimization of Computer Models,* John Wiley & Sons, Chichester.

Shannon, C.E. (1949): "Communication in Presence of Noise", *Proc. IRE,* **37**, 10–21.

Sinha, N.K. and Kuszta, B. (1983): *Modeling and Identification of Dynamic Systems,* Van Nostrand Reinhold Company, New York.

Sorenson, H.W. (1980): *Parameter Estimation – Principles and Problems,* Marcel Dekker Inc, New York.

Söderström, T. and Stoica, P.G. (1983): *Instrumental Variable Methods for System Identification,* Lecture notes in Control and Infornation Sciences (A.V. Balakrishnan and M. Thoma, eds), Springer-Verlag, Berlin.

Söderström, T. and Stoica, P.G. (1988): *System Identification,* Prentice-Hall, Englewood Cliffs, New Jersey.

Stoica, P., Eykhoff, P., Janssen, P. and Söderström, T. (1986): "Model-structure selection by cross-validation", *Int. J. Control,* **43**, 1841–1878.

Thoma, J.U. (1975): *Introduction to Bond Graphs and Their Applications,* Pergamon Press, Oxford.

Tulleken, H.J.A.F. (1990): "Generalized Binary Noise test-signal concept for improved identification experiment design, *Automatica,* **26**, 37–49.

Van den Bosch, P.P.J. and Lootsma, F.A. (1987): "Large scale electricity-production scheduling via nonlinear optimisation", *Journal of Optimisation: Theory and Applications,* **55**.

Van den Bosch, P.P.J. and Visser, H.R. (1990): "Simulation of state-events in power electronic devices", *Proc. 4th Intern. Conf. on Power Electronics and Variable Speed Drives,* London, 184-189.

Wahlberg, B. and Ljung, L. (1986): "Design variables for bias distribution in transfer function estimation", *IEEE Trans. Autom. Contr.,* **AC-32**, 134-144.

Zarrop, M.B. (1979): "Optimal experiment design for dynamic system identification", *Lect. not. in Contr. and Inf. Sciences,* **21**, Springer Verlag, Berlin.

Index

9 780849 391811